Monographs in Theoretical Computer Science
An EATCS Series

T0188838

Wil Michiels · Emile Aarts · Jan Korst

Theoretical Aspects of Local Search

With 70 Figures and 3 Tables

 Springer

Authors

Wil Michiels
Jan Korst
Philips Research
High Tech Campus 34
5656 AE Eindhoven
The Netherlands
wil.michiels@philips.com
jan.korst@philips.com

Emile Aarts
Philips Research
High Tech Campus 34
5656 AE Eindhoven
The Netherlands
and
Eindhoven University of Technology
P.O. Box 513
5600 MB Eindhoven
The Netherlands
emile.aarts@philips.com

Series Editors

Prof. Dr. Wilfried Brauer
Institut für Informatik der TUM
Boltzmannstr. 3
85748 Garching, Germany
brauer@informatik.tu-muenchen.de

Prof. Dr. Juraj Hromkovič
ETH Zentrum
Department of Computer Science
Swiss Federal Institute of Technology
8092 Zürich, Switzerland
juraj.hromkovic@inf.ethz.ch

Prof. Dr. Grzegorz Rozenberg
Leiden Institute of Advanced
Computer Science
University of Leiden
Niels Bohrweg 1
2333 CA Leiden, The Netherlands
rozenber@liacs.nl

Prof. Dr. Arto Salomaa
Turku Centre of
Computer Science
Lemminkäisenkatu 14 A
20520 Turku, Finland
asalomaa@utu.fi

ACM Computing Classification (1998): F.1, F.2, G.1, G.2, G.3, I.2.8.

ISSN 1431-2654

ISBN 978-3-642-07148-5 e-ISBN 978-3-540-35854-1

Springer is a part of Springer Science+Business Media
springer.com

© Springer-Verlag Berlin Heidelberg 2007
Softcover reprint of the hardcover 1st edition 2007

Cover Design: KünkelLopka, Heidelberg

Printed on acid-free paper 45/3100/YL 5 4 3 2 1 0

Preface

In 1958 Croes published a twenty page article in Operations Research describing a new method to handle traveling salesman problems based on the iterative use of simple edge-exchange mechanisms. This article can be seen as the seminal paper on local search, reporting for the first time the success of simple neighborhood-based iteration methods when applied to combinatorial optimization problems.

Half a century of practice reveals that local search has become quite extensively used, and evidence exists that it has become one of the most frequently and widely applied heuristic search methods in engineering, operations research, and design. By October 2006 the well-known search engine Google returned over twenty million entries upon the query 'local search'. Applications of local search are abundantly available in real life, and they are by no means limited to small-scale or insignificant toy problems. On the contrary, fields are known where local search provides a competitive edge to business and product engineering. For instance, in chip design, the use of tools applying local-search-based techniques for determining area-effective silicon layouts has resulted in smaller chips than those that could be obtained with other placement techniques, resulting in cost savings of several million euros on a yearly basis as a result of increased production yield. The use of local search techniques in logistics has enabled the computation of cost-effective routes for the transportation of packets that need to be shipped over a complex network of roads and railways. This improves throughput and shortens delivery time, which in turn leads to substantial cost savings.

Over the years, a huge number of papers have appeared on local search. Closer examination indicates that the vast majority of these papers is of an experimental nature. This observation reflects very well the position that local search has been taking now for almost half a century: it is primarily seen as a practical tool for which not much can be said from a theoretical point of view. Although we largely agree with this classification, it would be short sighted to consider the theory of local search as being insignificant.

Apart from a number of specific isolated results, one can identify three main topics of theoretical work relating to local search. In the first place, performance guarantees have been studied regarding the quality of solutions obtained with local search. Secondly, one has started to investigate the time complexity of local search. On the one hand, this has led to problem-specific results on the number of iterations required to reach a local optimum. On the other hand, a general theory on the time complexity of local search has been developed. This theory is the local search counterpart of the theory of NP-completeness. It can be used to prove the

intractability of finding a local optimum for some local search problems. As a third topic of theoretical work, we mention the asymptotic convergence of local search in the case where a probabilistic iteration mechanism is applied. Simulated annealing is probably the best known example of such a probabilistic variant of local search.

The aim of this book is to bring together the theoretical results relating to local search, following the tripartition given above. We do not intend to completely cover all theoretical results presented in the literature. Instead, we use a small set of classical combinatorial optimization problems as running examples, particularly the traveling salesman problem and machine scheduling. The results we prove for these problems can be considered typical for the results that can be found in this area. Since we restrict ourselves to theoretical results, this book should not be considered as a handbook on how to apply local search in practice. For example, we do not give any specific implementation details of the various local search metaheuristics, such as simulated annealing and tabu search. However, we aim at giving a better understanding of the fundamental behavior of local search.

The organization of this book is as follows. The introductory chapter positions local search within the field of combinatorial optimization, and presents basic definitions and terminology. Chapter 2 presents basic neighborhood functions for a number of classical combinatorial optimization problems. Chapter 3 illustrates the use of indirect representations for several machine scheduling problems. Chapter 4 discusses properties of neighborhood functions. It considers degree and diameter of neighborhood graphs as well as dominance relations between different neighborhood functions. In Chapter 5, we give several examples of performance guarantees for local minima of given neighborhood functions and we show, for some specific examples, that local optimality implies global optimality. Chapter 6 addresses the computational complexity of finding a local optimum for a given combinatorial optimization problem and a given neighborhood function. It discusses the complexity class PLS, and gives examples of PLS-complete problems and related PLS-reductions. Most of the results presented up to Chapter 6 relate to iterative improvement. In Chapter 7 we discuss metaheuristics that potentially give higher-quality solutions. These metaheuristics either allow non-improving moves, such as simulated annealing and tabu search, or use multiple runs, such as iterated local search and genetic local search. Finally, Chapter 8 deals with the asymptotic convergence of simulated annealing using the theory of Markov chains.

The presentation of this book is at a graduate level. The book can serve as an introductory textbook on the theory of local search. It contains many theorems and corresponding proofs as well as bibliographical notes and clarifying exercises. This makes the book quite suitable for students in mathematics, computer science, and electrical and industrial engineering. We recommend reading the chapters in the order in which they are presented. However, most of the chapters are reasonably self-contained, and the reader might like to immediately jump to a chapter of his or her interest. Nevertheless, we recommend the reader to at least read Chapter 1 for basic terminology. Some of the sections are marked with an asterisk, indicating that they provide additional material that can possibly be skipped.

Eindhoven, October 2006 Wil Michiels, Emile Aarts, Jan Korst

Contents

1

Introduction

In many different fields, including management science, computer science, and electrical and industrial engineering, we are confronted with combinatorial optimization problems. In such problems we are given a finite or countably infinite set of solutions from which we have to find one that minimizes or maximizes a given cost function. One of the best studied combinatorial optimization problems is the TRAVELING SALESMAN PROBLEM (TSP), which can be formulated as follows. Consider a salesman who wants to visit each city from a set of n cities exactly once and who wants to end up in the same city in which he started, where the distance between any two cities is given. In what order should the salesman visit the cities, such that the traveled distance is minimal? To see that this problem is indeed a combinatorial optimization problem, observe that it corresponds to picking a tour with minimum length from the finite set of all possible tours visiting the n cities.

As a representation of a tour we can use a permutation of the cities that is given by $\tau = (\tau(1), \tau(2), \ldots, \tau(n))$, where $\tau(i)$ denotes the ith city visited in the tour. As the number of tours, which equals the number of permutations, is factorial in the number of cities, it grows exponentially with the number of cities. More specifically, for one hundred cities the number of tours exceeds 10^{150}, which is larger than the estimated number of particles in the universe. So the solution space is extremely large, and the search for the optimum tour by evaluating them all is impracticable.

The formulation of TSP is characteristic of combinatorial optimization problems. Such problems, more specifically their solution space, can typically be formulated in terms of discrete structures, such as sequences, permutations, graphs, and partitions.

This use of discrete structures and the use of sophisticated algorithms that work on these structures, positions combinatorial optimization, which is the discipline that deals with combinatorial optimization problems, at the intersection of two well-developed scientific fields: discrete mathematics and computer science.

A major achievement in combinatorial optimization is the development of computational complexity theory. This theory formalizes the difference between easy and hard problems. A problem is called easy if it can be solved by a polynomial-time algorithm, where we say that an algorithm solves a problem if it always returns an optimal solution. A problem is called hard, formally referred to as NP-hard, if it is commonly believed that a polynomial-time algorithm that solves it does not exist. Many interesting combinatorial optimization problems, including TSP, have this property. We assume the reader to be acquainted with the basics of complexity theory. If not, the reader is referred to the standard work of Garey & Johnson [1979]. An overview of the discipline is given in Appendix B.

When confronted with an NP-hard combinatorial optimization problem, we have two options for tackling it. The first option is to aim for an optimal solution, despite the NP-hardness of the problem. One reason why this can still be feasible is that the instances at hand may have some special structure that makes the special case easy to solve. Another reason is that having an exponential running time does not necessarily mean that an algorithm based on enumeration, such as branch-and-bound, cannot be useful. It can, for instance, be adequate if one is interested in solving relatively small problem instances. Furthermore, the NP-hardness of a problem only indicates that (most probably) no algorithm exists with a *worst-case* polynomial running time. The average-case running time, however, can be much better than the worst-case running time.

A second option for tackling an NP-hard combinatorial optimization problem is to use a heuristic algorithm. Solutions found by such an algorithm are not necessarily optimal, but they are found within an acceptable amount of time. Hence, heuristic algorithms trade off optimality against computing time.

Heuristic algorithms can be classified into two categories: constructive algorithms and local search algorithms. A constructive algorithm generates a solution through a number of steps, where in each step the partial solution obtained so far is extended until in the last step a complete solution is obtained. The order in which the steps are carried out and the actions performed in each step are often strongly problem dependent. Local search algorithms, on the other hand, try to find high-quality solutions by searching through the solution space. More precisely, these algorithms start with an initial solution and then iteratively generate a new solution that is 'near' to the current solution. A neighborhood function defines for a given solution the solutions that are near to it. As mentioned above, solutions of many combinatorial optimization problems can be represented by discrete structures such as sequences, permutations, graphs, and partitions. Local search algorithms typically use these representations by defining neighborhood functions in terms of local rearrangements, such as swapping, moving, and/or replacing items, that may be applied to a representation to obtain a neighboring solution.

In this book, we focus on local search algorithms. In practice, excellent results have been obtained by using local search algorithms for a wide variety of problems. This has led to a growing interest in theoretical results concerning the approach in the past two decades. However, many problems are still open as a challenge for the interested reader. The material presented in this book can be subdivided into two main categories: on the one hand, theory and results that are problem independent, and on the other hand more dedicated results for several classical combinatorial optimization problems. Besides having their own merits, we hope the latter results help the reader to derive similar results for other problems.

We remark that this book should not be considered as a handbook on how to apply local search in practice. For instance, we do not give any specific details on how to implement various local search metaheuristics, such as simulated annealing and tabu search. The aim of this book is to give a better understanding of the fundamental behavior of local search by proving theoretical results.

1.1 Basics of Local Search

As mentioned, this book deals with combinatorial optimization problems. These problems can be formalized as follows.

Definition 1.1. An *instance* of a combinatorial optimization problem is a pair (S, f), where the solution space S is a finite or countably infinite set of solutions and the cost function f is a mapping $f : S \to \mathbb{R}$ that assigns a real value to each solution in S called the cost of the solution. □

Definition 1.2. A *combinatorial optimization problem* Π is specified by a set of problem instances and it is either a minimization or a maximization problem. The problem is to find for a given instance (S, f) a solution $s^* \in S$ that is *globally optimal*. For a minimization problem, this means that $f(s^*) \leq f(s)$ has to hold for all $s \in S$, and for a maximization problem it means that $f(s^*) \geq f(s)$ has to hold for all $s \in S$. If no confusion can arise, a globally optimal solution is simply called optimal. □

In this definition the adjective 'combinatorial' refers to the constraint that S has to be finite or countably infinite. If this constraint does not hold, then the problem is called an optimization problem. In a problem instance (S, f) of TSP, set S consists of all possible tours over a given number of cities and f defines for each tour in S its length. We note that a problem instance (S, f) is generally defined implicitly by using a compact data representation and not by giving the complete solution set S and the cost of each solution. A problem instance of TSP can, for instance, be defined by giving for each pair i, j of cities the distance from city i to j. The size of the representation, expressed as the number of bits required for storing it, is called the size of the problem instance. An algorithm is called a polynomial-time algorithm if its running time is polynomial in the size of a problem instance. Otherwise, it is called exponential.

We consider local search algorithms for tackling combinatorial optimization problems. The key feature of these algorithms is their neighborhood function. This

function specifies, for each solution, which solutions are in some sense near to it. The neighborhood function is generally defined in terms of small changes that may be applied to a solution to obtain a neighboring solution.

Definition 1.3. For an instance (S, f) of a combinatorial optimization problem, a *neighborhood function* is a mapping $N : S \to 2^S$, where 2^S denotes the powerset $\{V | V \subseteq S\}$. The neighborhood function specifies for each solution $s \in S$ a set $N(s) \subseteq S$, which is called the *neighborhood* of s. The cardinality of $N(s)$ is called the *neighborhood size* of s. We say that solution s' is a *neighbor* of s if $s' \in N(s)$. The neighborhood function N is said to be *symmetric* in the case that we have $s' \in N(s)$ if and only if $s \in N(s')$. \square

Throughout this section, we let S and f define a problem instance of an arbitrary combinatorial optimization problem and we let N be an arbitrary neighborhood function for S. Furthermore, without loss of generality we only consider minimization problems. A maximization problem can be transformed into a minimization problem by reversing the sign of cost function f.

A local search algorithm starts with an initial solution that is constructed by some heuristic algorithm. Next, the local search algorithm searches through the solution space by continually moving from a solution to one of its neighbors. This process can be modeled as a walk through the neighborhood graph.

Definition 1.4. The *neighborhood graph* of an instance (S, f) of a combinatorial optimization problem and an accompanying neighborhood function N is a directed node-weighted graph $G = (V, E)$. The node set V is given by the set S of solutions, and the arc set E is defined such that $(i, j) \in E$ if and only if $j \in N(i)$. Furthermore, we define the weight of a node as the cost of the corresponding solution. If the neighborhood function is symmetric, then the directed graph can be simplified to an undirected graph by replacing arcs (i, j) and (j, i) by a single edge $\{i, j\}$. \square

We note that the neighborhood functions used in practice are almost always symmetric. We assume the reader to be familiar with basic graph theory. An overview of the basic graph theory concepts used in this book is presented in Appendix A.

Definition 1.5. A solution j is *reachable* from solution i if the neighborhood graph G contains a path from i to j. This means that a sequence of solutions s_1, s_2, \ldots, s_k exists with $k \geq 1$ such that $s_1 = i$, $s_k = j$, and $s_{l+1} \in N(s_l)$ with $1 \leq l < k$. \square

Definition 1.6. A neighborhood graph is *strongly connected* if, for each pair i, j of solutions, j is reachable from i. A neighborhood graph is *weakly optimally connected* if, for each solution i, it contains a path from i to an optimal solution. \square

Different strategies have been proposed for walking through a neighborhood graph. The most obvious strategy is used by the *iterative improvement algorithm*, also known as the *hill climbing algorithm*. This is the basic local search algorithm. In each iteration, the algorithm searches in the neighborhood of the current solution for a solution with better cost. If such a solution is found, it replaces the current

solution. Otherwise, the algorithm stops and returns the current locally optimal solution.

Definition 1.7. A solution $\hat{s} \in S$ is called *locally optimal* with respect to N or N-*optimal* if $f(\hat{s}) \leq f(s)$ for all $s \in N(\hat{s})$. □

In this book, a star as superscript indicates global optimality and a hat indicates local optimality. Hence, s^* is a globally optimal solution, whereas \hat{s} is locally, but not necessarily globally optimal.

We note that if we have two different neighborhood functions N_1 and N_2 for the same problem instance and if N_1 is dominated by N_2, as defined below, then each local (global) optimum for N_2 is also a local (global) optimum for N_1.

Definition 1.8. Let N_1 and N_2 be two different neighborhood functions for the same instance (S, f) of a combinatorial optimization problem. If for all solutions $s \in S$ we have $N_1(s) \subseteq N_2(s)$, then we say that N_2 *dominates* N_1. □

Instead of being modeled as a walk through a neighborhood graph, an execution of iterative improvement can be modeled more precisely as a walk through a transition graph.

Definition 1.9. The *transition graph* of an instance of a combinatorial optimization problem and an accompanying neighborhood function is a directed, acyclic subgraph of their neighborhood graph G. It is obtained from G by deleting all arcs (i, j) for which it holds that the cost of solution j is worse than or equal to the cost of solution i. □

Note that a solution is locally optimal if and only if it has outdegree zero in the transition graph. The rule used by iterative improvement to select a neighboring solution in the case that a solution has multiple neighbors with a better cost is called the *pivoting rule*. Well-known pivoting rules are *first improvement* and *best improvement*. In first improvement we generate neighbors and we accept the first solution encountered with better cost, where the neighbors can be generated randomly or in some specified order. When using best improvement, we replace s by a best solution in its neighborhood, provided that it has better cost. Figure 1.1 gives the iterative improvement algorithm in pseudo-code. If the neighborhood function is exact, then the algorithm is guaranteed to give a globally optimal solution.

Definition 1.10. A neighborhood function is called *exact* if each local optimum is also a global optimum. □

Example 1.1. To illustrate the definitions given above, consider the problem instance (S, f) and the (symmetric) neighborhood function N defined in Table 1.1. Figure 1.2 depicts the corresponding neighborhood graph. The figure also shows the corresponding transition graph, which can be derived from the neighborhood graph. The problem instance has two global optima, namely solutions 3 and 6. Furthermore, solution 1 is a local optimum. This implies that the neighborhood function

algorithm Iterative Improvement

begin
 $s :=$ some initial solution;
 repeat
 generate an $s' \in N(s)$;
 if $f(s') < f(s)$ **then** $s := s'$;
 until $f(s') \geq f(s)$ for all $s' \in N(s)$;
end;

Figure 1.1. Iterative improvement algorithm for a minimization problem. The first-improvement pivoting rule generates neighbors from $N(s)$ at random or in some specified order. The best-improvement pivoting rule generates a neighbor from $N(s)$ with lowest cost.

we consider is not exact. To make it exact, it suffices to change the neighborhood function such that solutions 1 and 3 are neighbors of each other.

The neighborhood graph of Figure 1.2 contains two disconnected components: the subgraph G_1 induced by solutions 1, 2, 3, and 4 and the subgraph G_2 induced by solutions 5, 6, and 7. Solutions from G_1 are not reachable from solutions from G_2 and vice versa. Hence, the neighborhood graph of Figure 1.2 is not strongly connected. However, as G_1 and G_2 both contain a global optimum it is weakly optimally connected. □

One often uses the metaphor of walking in a mountainous region to get an intuitive idea of local search. Consider an instance of a minimization problem. If we define the height of a solution as its cost, then walking through a neighborhood graph to find a global optimum can be seen as walking on a three-dimensional surface to find the lowest point. This looks like walking in a dense fog over the surface of the earth with its mountains and valleys, where we are looking for the lowest point and we cannot see farther than one step ahead. Furthermore, a local optimum corresponds to the lowest point in a valley and applying iterative improvement means that we only allow downhill moves.

A major drawback of applying iterative improvement is that it may get trapped in poor local optima. One way to tackle this problem is to find a better, possibly more complex, neighborhood function. Alternatively, one can allow non-improving moves or perform multiple runs of iterative improvement. The latter two approaches are frequently used in practice and often with success. The most popular local search algorithms based on them are discussed in Chapter 7.

The performance of iterative improvement is determined not only by the quality of the local optima, but also by the time it takes to reach a local optimum. The time required to reach a local optimum is determined by two aspects: the time needed for moving from one solution to the next and the number of solutions that are visited before one arrives at a local optimum. To keep the first part small, it is important

Table 1.1. Problem instance (S, f) with $S = \{1, 2, \ldots, 7\}$ and neighborhood function N discussed in Example 1.1.

solution	$f(s)$	$N(s)$
1	2	$\{2\}$
2	3	$\{1,3\}$
3	1	$\{2,4\}$
4	2	$\{3\}$

solution	$f(s)$	$N(s)$
5	2	$\{6\}$
6	1	$\{5,7\}$
7	2	$\{6\}$

that the outdegree of the nodes in the neighborhood graph is not too large, i.e., the cardinality of the neighborhoods of solutions may not be too large. Consider, for instance, the extreme case in which the neighborhood graph is a complete graph. This means that for each $i, j \in S$, we have $i \in N(j)$. We then have high-quality local optima because the neighborhood function is exact. However, testing for local optimality and finding a better neighbor in the case that a solution is not locally optimal comes down to solving the original problem.

To keep the number of iterations to reach a local optimum small, it is important that the transition graph has a small 'potential'.

Definition 1.11. Let T be a transition graph, and let \hat{V} be the nodes in T with out-degree zero, i.e., the nodes that correspond to local optima. The *potential* of a node v is defined as the minimum distance of v to a node in \hat{V}, where the distance from node i to j is defined as the minimum number of arcs in a connecting path from i to j. The potential of transition graph T is the maximum potential over all its nodes. □

The potential of a transition graph gives a lower bound on the number of iterations that are maximally required by iterative improvement if it can select an arbitrary solution as the starting solution. The lower bound is tight whenever an optimal pivoting rule is used, where optimal means that a shortest path to a locally optimal solution is chosen.

As mentioned, local search algorithms that admit non-improving moves are often used in practice to overcome the problem of ending up in poor local optima. In the remainder of this section, we consider such algorithms and discuss some related issues. First, we observe that they can still result in low-quality solutions if a neighborhood graph consists of several disconnected components, i.e., if the solution space can be partitioned into subsets V_1, V_2, \ldots, V_m, such that from solution $i \in V_j$ only other solutions from V_j can be reached. If in that case we start with a solution from a subset V_j with only low-quality solutions, then the local search algorithm still has a poor performance. Obviously, this problem does not occur if we require the neighborhood graph to be strongly connected. However, when we start with an initial solution i, we are actually not interested in whether all solu-

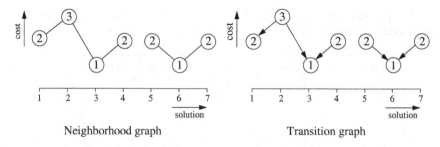

Figure 1.2. Neighborhood graph and transition graph of the problem instance and neighborhood function defined in Table 1.1. The number written inside a node defines its weight, i.e., the cost of the solution it represents.

tions are reachable. We only want to know whether we can reach a global optimum. Hence, the condition that the neighborhood graph must be strongly connected can be weakened to that it must be weakly optimally connected.

We do not only want that a global optimum is reachable. We also want it to be reachable within a reasonable number of steps. This is realized in a strongly connected neighborhood graph if its diameter is not too large.

Definition 1.12. The *diameter* of a neighborhood graph is the maximum distance between any pair of nodes, where the distance is as defined in Definition 1.11. If the neighborhood graph is not strongly connected, then its diameter is defined to be infinitely large. □

Although the performance of a local search algorithm generally improves by allowing deteriorating moves, it should not be too eager to carry out these moves in order to prevent it from carrying out a random search. As a result, the difficulty in reaching a solution j from a solution i via a path p strongly depends on the height of the path.

Definition 1.13. Let $p = (s_1, s_2, \ldots, s_k)$ be a path in a neighborhood graph, where s_i is the ith node in the path. Then the *height* of path p is given by the maximum label of any of its nodes, i.e., by $\max_{1 \le i \le k} f(s_i)$. □

Using this definition, we can define the depth of a local optimum to express how hard it is to escape it.

Definition 1.14. Let $\hat{s} \in S$ be a local optimum. The *depth* of \hat{s} is defined as the minimum height of a path p from \hat{s} to a solution s with $f(s) < f(\hat{s})$. If such a solution s does not exist, then the depth of \hat{s} is ∞. □

To obtain an effective local search algorithm, it is important that the solution space does not contain large *plateaus*, where a plateau refers to a part of the solution space in which all solutions have (approximately) the same value. On a plateau it is difficult for a local search algorithm to guide itself to better solutions as all

directions look the same. Another problem arising on plateaus is *cycling*. Cycling means that the algorithm revisits solutions over and over again. If, for instance, we have a plateau containing two local optima, s and s', that are neighbors of each other, have the same cost, and are surrounded by solutions with very high cost, then a local search algorithm will tend to alternate between s and s'. To avoid cycling a local search algorithm can remember the last visited solutions or fingerprints of these solutions.

1.2 Outline of the Book

The remainder of this book is organized as follows. In the next chapter we present some classical combinatorial optimization problems with accompanying neighborhood functions. Besides illustrating how the use of different solution representations may give rise to different neighborhood functions, these are the problems and neighborhood functions for which we prove problem-dependent results in this book.

Neighborhood function can be based on representations that directly define a solution and on representations that indirectly define a solution. In Chapter 3 we show the use of indirect representations. In Chapter 4 we derive properties of neighborhood functions, such as the diameter and connectivity of their corresponding neighborhood graph. Performance guarantees for neighborhood functions, i.e., for their local optima, are proved in Chapter 5. In this chapter we also prove the negative result that if it is NP-hard to find an R-approximate solution for some problem, then the problem does not admit an efficient neighborhood function with a performance bound of R. While Chapter 5 relates to the quality of local optima, Chapter 6 is concerned with the time complexity of finding a local optimum. This includes a formal theory that is similar to the theory of NP-completeness.

Chapter 7 elaborates on several different techniques to overcome the problem that iterative improvement stops at the first local optimum it encounters. One of the techniques presented is simulated annealing, for which we prove in Chapter 8 that, under some mild conditions, it converges asymptotically to the set of globally optimal solutions.

1.3 Bibliographical Notes

Classical textbooks on combinatorial optimization are Cook, Cunningham, Pulleyblank & Schrijver [1998], Korte & Vygen [2002], Lawler [1976], Nemhauser & Wolsey [1988], and Papadimitriou & Steiglitz [1982]. All these textbook have been written for a more advanced audience. Foulds [1984] presents a more introductory textbook written at the undergraduate level. Beale [1988] presents a more general introduction to optimization with a chapter on combinatorial optimization. Recently, Schrijver [2003] published a three-volume textbook that provides a quite elaborate and in-depth presentation of the field with a wealth of references to the existing literature. Annotated bibliographies on combinatorial optimization are given by O'Heigeartaigh, Lenstra & Rinnooy Kan [1985] and Dell'Amico, Maffioli & Martello [1997].

Aarts & Lenstra [1997] present an overview of both theoretical and practical aspects of local search. Hoos & Stützle [2004] present a book that serves as a practical guide to the application of local search algorithms and its many variants. A gentle introduction to local search is given by Michalewicz & Fogel [2000]. Overviews on local search algorithms are given by Glover & Kochenberger [2003], Reeves [1993], and Ribeiro & Hansen [2002]. An annotated bibliography on local search is given by Aarts & Verhoeven [1997].

1.4 Exercises

1. The set $A = \{1,3,6,9\}$ of integers is given. We define the cost of a sequence τ of the numbers in A as $f(\tau) = \sum_{i=1}^{4} i\tau(i)$, where each number of A occurs exactly once in τ and where $\tau(i)$ denotes the ith integer in τ.

 We consider the problem of finding a sequence τ over A that minimizes $f(\tau)$. This problem is equivalent to finding a descending ordering of the numbers in A.

 a) Give the solution space of the described problem instance.

 b) Give the cost of each solution/sequence.

 Consider the swap neighborhood function in which sequence τ' is a neighbor of sequence τ if and only if τ' can be obtained from τ by changing the order of two adjacent numbers.

 c) Give the neighborhood graph for the swap neighborhood function.

 d) Give the transition graph for the swap neighborhood function.

 e) Verify that the swap neighborhood function is exact.

 f) Determine the depth of the local optima.

2. Consider the iterative improvement algorithm described in Figure 1.1. We apply this algorithm to the problem instance of Exercise 1, where an initial solution is selected uniformly at random. Show that applying the best-improvement pivoting rule results in a smaller expected number of iterations than applying the first-improvement pivoting rule, where the first-improvement pivoting rule generates neighbors uniformly at random.

3. We defined iterative improvement such that a neighbor is only accepted if it has strictly better cost. Give a reason to not accept a neighbor that has equal cost.

2

Basic Examples

We discuss some classical combinatorial optimization problems and present neighborhood functions for them in Sections 2.1 to 2.5. Furthermore, we discuss an interesting link between local search in combinatorial optimization and neural computing in Section 2.6.

2.1 Traveling Salesman Problem

One of the most studied combinatorial optimization problems is the TRAVELING SALESMAN PROBLEM. In its general form the problem is defined as follows.

Definition 2.1 [TRAVELING SALESMAN PROBLEM (TSP)]. Given are a set of n cities denoted by $C = \{1, 2, \ldots, n\}$ and an $n \times n$ matrix d, where $d_{ij} \in \mathbb{N}^+$ defines the distance from city i to city j. Find a tour with minimum total tour length, i.e., find a permutation τ of C that minimizes

$$\sum_{i=1}^{n-1} d_{\tau(i),\tau(i+1)} + d_{\tau(n),\tau(1)}.$$

□

In this definition permutation τ specifies the tour in which $\tau(i)$ is the ith city that is visited. Correspondingly, we write a permutation τ as $\tau = (\tau(1), \tau(2), \ldots, \tau(n))$. This must not be confused with the alternative cyclic notation, where (i_1, i_2, \ldots, i_n) means that i_1 is the image of i_n and that i_{j+1} is the image of i_j for $1 \leq j < n$. This cyclic notation is also frequently used in the literature to define tours, but then $\tau(i_j)$

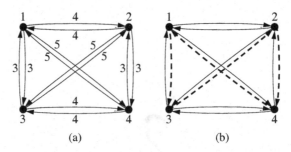

Figure 2.1. (a) Complete directed arc-weighted graph corresponding to the problem instance discussed in Example 2.1. (b) Hamiltonian cycle corresponding to the tour $\tau = (1,3,2,4)$.

specifies the city visited directly after city i_j. The former notation is used for all permutations occurring in this book because permutations will in most cases define arrangements, i.e., mappings from positions to some kind of items.

TSP can be viewed as finding a minimum weighted Hamiltonian cycle in a complete, directed, arc-weighted graph with n nodes, where the weight of an arc (i, j) is given by the distance d_{ij} from city i to city j. Correspondingly, we use terminology such as an arc of a tour. If in this book we write $(\tau(j), \tau(j+1))$ to denote an arc of a tour τ, then we mean arc $(\tau(n), \tau(0))$ in the case where $j = n$.

Example 2.1. Consider the problem instance of TSP in which we are given four cities that represent the corners of a four-by-three rectangle. This means that we are given the cities 1, 2, 3, and 4 and the distance matrix

$$d = \begin{pmatrix} 0 & 4 & 3 & 5 \\ 4 & 0 & 5 & 3 \\ 3 & 5 & 0 & 4 \\ 5 & 3 & 4 & 0 \end{pmatrix}.$$

Then a one-to-one correspondence exists between tours for this problem instance and Hamiltonian cycles in the complete, directed, arc-weighted graph depicted in Figure 2.1(a). For instance, the tour $\tau = (1,3,2,4)$ that starts in city 1 and, before returning to it, successively visits cities 3, 2, and 4 corresponds to the Hamiltonian cycle depicted in Figure 2.1(b). The length of the tour is given by the total weight of the arcs in the Hamiltonian cycle, which in this case is 16. □

For a problem instance of TSP, the size of the solution space is given by the number of different tours. As a tour is determined by the order in which it visits the cities and not by the city in which it starts, this number is given by the number of permutations of C divided by n. Hence, the size of the solution space is $(n-1)!$.

TSP is one of the most difficult problems in the class of NP-hard problems. Even finding a tour whose length is guaranteed to be within $2^{p(n)}$ times the optimal length for some arbitrary fixed polynomial p is already NP-hard. However, in many

applications we are only interested in a special case of TSP, where the distance matrix satisfies some natural conditions. For example, the distance matrix is often symmetric, which means that $d_{ij} = d_{ji}$ holds for each pair $i, j \in C$. In this case, we can consider the Hamiltonian cycle that represents a tour to be undirected. Note that this implies that the size of the solution space halves.

Besides being symmetric, the distance matrix often also satisfies the triangle inequality. This constraint states that moving from city i to j via k is not shorter than moving from i directly to j. Formally, this means that $d_{ij} \leq d_{ik} + d_{kj}$ holds for all i, j, k. The triangle inequality is, for instance, satisfied if the cities are points in an m-dimensional space for some $m \geq 1$ and if we consider the Euclidean distance matrix, which means that for $i = (x_1, x_2, \ldots, x_m)$ and $j = (y_1, y_2, \ldots, y_m)$ the distance from city i to j is given by

$$d_{ij} = \sqrt{|x_1 - y_1|^2 + |x_2 - y_2|^2 + \cdots + |x_m - y_m|^2}.$$

For two-dimensional space, the equation above defines the Euclidean distance $d_{ij} = \sqrt{|x_1 - y_1|^2 + |x_2 - y_2|^2}$.

The variant of TSP in which the distance matrix is constrained to be symmetric is called SYMMETRIC TSP. If, in addition, the distance matrix has to satisfy the triangle inequality, then the problem is called METRIC TSP, and the special case of METRIC TSP in which we assume that the cities lie in two-dimensional space and that the distance matrix is Euclidean is referred to as EUCLIDEAN TSP. Note that Example 2.1 discusses a problem instance of EUCLIDEAN TSP. When formulating a result we use the fact that we need not explicitly give each variant of TSP for which it holds. Positive results translate directly to more restricted problems and negative results to less restricted problems.

Since computers cannot cope with real numbers of arbitrary precision, we defined TSP such that all distances have integer values. For TSP, SYMMETRIC TSP, and METRIC TSP this constraint is reasonable. If an application implies a real-valued distance d_{ij}, then the effect of increasing it to the smallest integer larger than d_{ij}, i.e., to $\lceil d_{ij} \rceil$, can be made arbitrarily small by choosing a finer granularity. For EUCLIDEAN TSP this trick causes a problem because if the distance d_{ij} from city $i = (x_1, x_2)$ to $j = (y_1, y_2)$ is increased to

$$\lceil d_{ij} \rceil = \left\lceil \sqrt{|x_1 - y_1|^2 + |x_2 - y_2|^2} \right\rceil, \tag{2.1}$$

then it is no longer Euclidean. A straightforward way to solve this problem is to redefine EUCLIDEAN TSP such that the distance from city i to j is given by (2.1). To simplify the discussion, we nevertheless assume throughout this book the original definition $d_{ij} = \sqrt{|x_1 - y_1|^2 + |x_2 - y_2|^2}$, which means that we allow real-valued distances. It can be verified that, by choosing a fine enough granularity, all results we derive also hold for this more properly defined problem.

All four proposed variants of TSP are strongly NP-hard. Hence, from a complexity point of view, they are all equally hard to *solve*. However, the theorem below states that they are not all equally hard to *approximate*. More precisely, the theorem

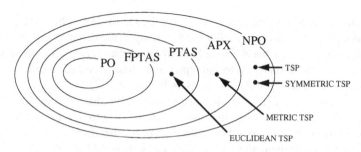

Figure 2.2. Position of the different variants of TSP in the existing approximability classes in the case that P \neq NP. The approximability classes are discussed in Appendix B.

states that the variants are positioned in the existing approximability classes as indicated in Figure 2.2. If one is not familiar with the terminology used in the theorem and in the figure, we refer to Appendix B. The theorem gives a survey of complexity results scattered in the literature. In the proof of the theorem, we restrict ourselves to giving references to where the results are proved.

Theorem 2.1. *For the different variants of* TSP, *we have the following complexity results.*

- *Both for* TSP *and* SYMMETRIC TSP, *even finding a tour with a length at most* $2^{p(n)}$ *times the optimal tour length is NP-hard, where p is an arbitrary fixed polynomial.*

- *For* METRIC TSP, *a polynomial-time* $\frac{3}{2}$-*approximation algorithm exists, but finding a tour with length at most* $\frac{220}{219}$ *times the optimal tour length is NP-hard.*

- *For* EUCLIDEAN TSP, *a PTAS exists, but, unless P = NP, the problem does not admit a FPTAS.*

Proof. The result for SYMMETRIC TSP, which also implies the result for standard TSP, is proved by Sahni & Gonzalez [1976].

For METRIC TSP, Christofides [1976] proposes a polynomial-time algorithm for which he proves that it has a performance bound of $\frac{3}{2}$, and Cornuejols & Nemhauser [1978] show that this performance bound is tight. Currently, no polynomial-time algorithm exists for which a better worst-case performance bound has been derived. Furthermore, Papadimitriou & Vempala [2000] prove that unless P = NP, METRIC TSP cannot be approximated in polynomial time within $\frac{220}{219}$ times the optimum.

For EUCLIDEAN TSP, Arora [1998] derives a PTAS. This is the best we can hope for as Papadimitriou [1977] and Garey, Graham & Johnson [1976] prove that the problem is strongly NP-hard, which implies that it does not admit a FPTAS, unless P = NP [Ausiello et al., 1999]. □

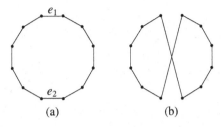

Figure 2.3. (a) Original tour. (b) Only tour that can be obtained by replacing edges e_1 and e_2 by two new edges.

k-**change.** A well-known local search algorithm for TSP is *k*-Opt, which is iterative improvement based on the *k*-change neighborhood function. A tour is a neighbor of a tour with respect to this neighborhood function if it can be obtained by performing a *k*-change. A *k*-change operation deletes $k' \leq k$ edges and replaces them with k' new edges, such that we again get a feasible tour, i.e., a Hamiltonian cycle. In practice, one generally uses $k = 2, 3$. Note that if we consider TSP in its general form, in which case the distance matrix does not have to be symmetric, then a tour contains arcs instead of edges. We then obtain a *k*-change neighbor of a tour τ by first undirecting tour τ, by next performing a *k*-change, and by finally choosing one of the two possible directions for the derived tour.

As an example, let $k = 2$ and consider a problem instance of SYMMETRIC TSP with twelve cities. Furthermore, let τ be an arbitrary tour. In Figure 2.3(a) tour τ is visualized by a circle containing the cities in the order in which they are visited. The real distances between cities are not related to the distances in the figure. This is a commonly used approach for visualizing a tour. As mentioned, a 2-change removes at most 2 edges from the tour and replaces them by new ones. Obviously, we cannot replace a single edge by a new one. Hence, the only non-trivial 2-changes are those in which two edges e_1 and e_2 are replaced by two new edges. For some choice of e_1 and e_2, Figure 2.3(b) shows the only way to do this in such a way that a feasible tour is obtained. It can be verified that if e_1 and e_2 are successive edges, then it is not possible to construct a feasible tour by adding two new edges. Hence, the neighborhood of tour τ contains tour τ, which is obtained by replacing zero edges, and one neighbor for each choice of two non-successive edges e_1 and e_2. This implies that tour τ has $1 + n(n-3)/2 = 55$ neighbors in total. Stated more generally, we find that for any problem instance of SYMMETRIC TSP and for any tour τ, the 2-change neighborhood size is $1 + n(n-3)/2 = \Theta(n^2)$. Note that if the distance matrix need not be symmetric, then we also have to choose the direction of the tour, which implies that the neighborhood size increases by a factor of two to $2 + n(n-3)$. By inspecting Figure 2.3, it can be verified that applying a 2-change to a tour corresponds to reversing some segment of the tour. An example of a 3-change is given in Figure 2.4.

For SYMMETRIC TSP an advantage of the *k*-change neighborhood function is that if we know the length l of a tour τ and we want to know the length l' of a

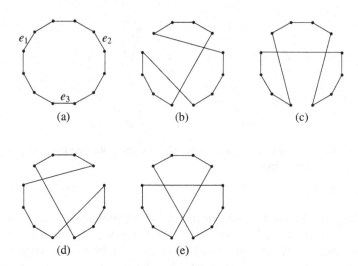

Figure 2.4. (a) Original tour. (b)-(e) Tours obtained by performing a 3-change on the edges e_1, e_2, and e_3.

neighboring tour τ' of τ, then we do not have to derive l' from scratch, which would take $\Theta(n)$ time, but we can use the value l to derive it in constant time. More specifically, we have

$$l' = l - \sum_{i=1}^{k'} d(e_i) + \sum_{i=1}^{k'} d(e_i'),$$

where $d(e)$ denotes d_{ij} for an edge $e = \{i, j\}$ and where the edges $e_1, e_2, \ldots, e_{k'}$ with $k' \leq k$ are removed and the edges $e_1, e_2, \ldots, e_{k'}$ are added by the k-change that turns τ into τ'. As the 2-change neighborhood size is $\Theta(n^2)$, as we derived above, this implies that for the 2-change neighborhood function a single iteration of iterative improvement can be implemented to run in $\mathcal{O}(n^2)$ time.

Example 2.2. Figure 2.5 depicts the five western islands of the Canary Islands. A man living in Maspalomas wants to visit the cities Santa Cruz de la Palma, San Sebastián, Playa de las Americas, and Valverde by boat. To save fuel, he is interested in finding the shortest tour that starts and ends in Maspalomas and that visits each of these cities exactly once. This means that he wants to solve a TSP instance with five cities. The five cities are all reachable by boat, and the distances in kilometers between the cities are given by the symmetric distance matrix

$$d = \begin{pmatrix} 0 & 233 & 152 & 118 & 227 \\ 223 & 0 & 95 & 115 & 90 \\ 152 & 95 & 0 & 30 & 93 \\ 118 & 115 & 30 & 0 & 115 \\ 227 & 90 & 93 & 115 & 0 \end{pmatrix}.$$

Entry d_{ij} gives the distance from city i to city j.

Figure 2.5. Five cities in the Canary Islands for which we want to find the shortest tour.

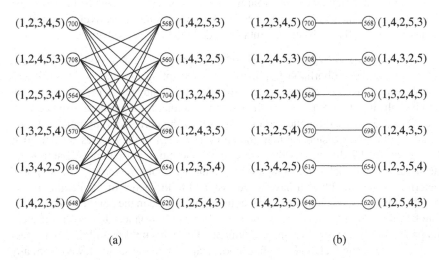

(a) (b)

Figure 2.6. (a) Neighborhood graph of Example 2.2, where we omit loops. Inside a node, we write its weight, i.e., the cost of the corresponding tour. (b) The same graph, where only the edges are drawn in which this graph differs from the complete bipartite graph.

Consider the 2-change neighborhood function for this problem instance. If we leave loops out of consideration, the neighborhood graph is given by the bipartite graph $B = (V, U, E)$ depicted in Figure 2.6(a). In the figure we use the fact that, as the direction of the tour is irrelevant, we only have to consider tours in which the cities 1, 2, and 5 are visited in increasing order. The neighborhood graph is a complete bipartite graph, except that, for each node $v \in V$, a unique node $g(v) \in U$ exists, such that the edge $\{v, g(v)\}$ is missing; see Figure 2.6(b). Obviously, the minimum length of a path between two solutions i and j, $i \neq j$, expressed in the number of edges, is two if i and j both belong to either subset V or subset U. If, on the other hand, $i \in V$ and $j \in U$ (or vice versa), then this length is three if $j = g(i)$ ($i = g(j)$) and

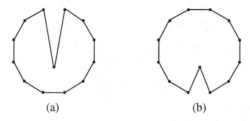

(a) (b)

Figure 2.7. (a) Original tour. (b) Tour obtained by a node-insertion move.

one otherwise. As a consequence, the diameter of the neighborhood graph is three. Furthermore, it follows from the graph that the neighborhood function is exact for this example and that the optimal tour has a cost of 560 and is given by $(1,4,3,2,5)$. This means that the shortest tour takes 560 kilometers and successively visits Playa de las Americas, San Sebastian, Santa Cruz de la Palma, and Valverde. □

The example shows that for five cities, the 2-change neighborhood function has a well-structured neighborhood graph. This is no exception, as can be seen as follows. The construction of a k-change neighborhood graph for any problem instance of TSP can be split into two steps. In the first step we define, for a problem instance with the cities $1,2,\ldots,n$, the node set and edge set of the corresponding neighborhood graph. In the second step we map the cities $1,2,\ldots,n$ to the n cities we consider. For Example 2.2 this means that we number the cities Maspalomas, Santa Cruz de la Palma, San Sebastian, Playa de las Americas, and Valverde. The mapping determines the weights that have to be assigned to the nodes of the neighborhood graph. Obviously, by choosing a proper mapping, a node in the neighborhood graph can be related to any tour for the considered problem instance. However, different mappings imply different weight assignments. From this it follows that, with respect to its structure, the k-change neighborhood graph looks the same viewed from any node, i.e., the neighborhood graph is isotropic.

Definition 2.2. A permutation π of the node set of graph $G = (V,E)$ is called an *automorphism* if for all $i, j \in V$, we have $\{i, j\} \in E$ if and only if $\{\pi(i), \pi(j)\} \in E$. □

Definition 2.3. A graph $G = (V,E)$ is called *isotropic* if for all $i, j \in V$ an automorphism π exists that maps i to j, i.e., for which $\pi(i) = j$. □

Node insertion. A very simple alternative neighborhood function for TSP is node insertion. For this neighborhood function the neighbors of a tour τ are obtained by deleting one city in τ and inserting it at another position in the tour. Figure 2.7 shows a node-insertion move for a problem instance of SYMMETRIC TSP. Again, the figure does not give an exact picture of the problem instance, but only shows the order in which cities are visited.

An obvious generalization of node insertion is the neighborhood function that, instead of inserting a single city, inserts a segment of successive cities between two

neighboring cities elsewhere in the tour. The Or neighborhood function is defined in this way, where the segment may consist of at most three cities. Because an Or move can be written as a 3-change (see Exercise 2 in Chapter 4), the Or neighborhood function can also be viewed as a restricted 3-change neighborhood function instead of a generalized version of node insertion. This illustrates two techniques for fine tuning a neighborhood function to one's problem and resources: generalization, which results in larger, more powerful neighborhoods that are, however, harder to evaluate, and restriction, which results in smaller, less powerful neighborhoods that are easier to evaluate.

2.2 Machine Scheduling

Machine scheduling is a generic name for a class of problems that are concerned with the assignment of n jobs to m machines to optimize a given criterion. The jobs may consist of multiple operations, and the machines can process at most one operation at a time. A schedule specifies for each operation on which machine and in which time interval it is processed. The set of jobs is given by $J = \{1, 2, \ldots, n\}$ and the set of machines by $M = \{1, 2, \ldots, m\}$. Each job may have a due date. The due date is the time at which a job has to be completed. A job may be completed after its due date, but this incurs a penalty. A particular machine scheduling problem is specified by the three-field classification $\alpha|\beta|\gamma$, where α defines the machine environment, β the job characteristics, and γ the optimality criterion. We now discuss these three characteristics in more detail.

Machine environment. The machine environment is given by $\alpha = \alpha_1\alpha_2$, where α_1 gives the type of the machines and α_2 the number of machines. If we have $\alpha_2 = \circ$, where \circ represents the empty symbol, then the number of machines is variable. If, on the other hand, α_2 is given by a positive integer m, then we have a fixed number m of machines. Next, consider α_1, which is an element of $\{\circ, P, Q, R, F, O, J\}$. For $\alpha_1 \in \{\circ, P, Q, R\}$ we have a single-stage system. This means that each job j consists of only one operation, which can be processed on any machine. The time required for processing job j on machine i is denoted by $p_{ij} \in \mathbb{N}^+$. If we only have one machine or if p_{ij} is independent of i, then we leave out the dependency on i in the notation and simply write p_j.

- $\alpha_1 = \circ$. We have only one machine on which all jobs have to be scheduled. In this case $\alpha_2 = 1$ has to hold.

- $\alpha_1 = P$. We consider identical parallel machines. This means that the time required for processing a job is equal on all machines.

- $\alpha_1 = Q$. We consider uniform parallel machines. Each job j has a fixed processing requirement p_j, but the machines can have different speeds. This means that the time required for processing job j on machine i is given by $p_{ij} = p_j/s_i$, where s_i denotes the speed of machine i. If each machine has speed 1, then we arrive at the previous case.

- $\alpha_1 = R$. We consider unrelated parallel machines. The processing requirements p_{ij} may be arbitrary positive numbers.

Note that each case generalizes the previous one.

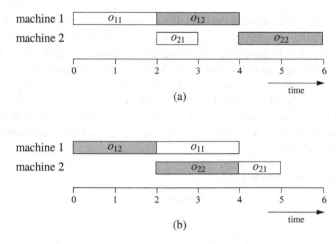

(a)

(b)

Figure 2.8. Gantt chart for visualizing the schedules of Example 2.3. The execution of an operation is represented by a box, the length of which is given by its processing requirement.

For $\alpha_1 = F, O$, and J, we have a flow shop, open shop, and job shop, respectively. In these cases we have a multi-stage system in which a job can consist of multiple operations. In a schedule the time intervals at which operations of the same job are being processed may not overlap.

Both in a flow shop and an open shop, we have that each job j is defined by a set $\{o_{1j}, o_{2j}, \ldots, o_{mj}\}$ of m operations. Operation o_{ij} has to be processed on machine i and its processing takes $p_{ij} \in \mathbb{N}^+$ time units. The difference between a flow shop and an open shop is that in the former case the operations of a job are processed in increasing order, i.e., on the machines $1, 2, \ldots, m$ in that order, whereas in an open shop choosing the machine routing for each job is part of the problem.

In a job shop, job j consists of m_j operations $\{o_{1j}, o_{2j}, \ldots, o_{m_jj}\}$, which have to be processed in that order. Operation o_{ij} has to be processed on a given machine μ_{ij} for $p_{ij} \in \mathbb{N}^+$ time units. Note that the i in p_{ij} does not refer to a machine, as is the case for the other choices of α_1. It may be the case that different operations of the same job have to be assigned to the same machine. This is, however, not true if the operations are successive. Hence, $\mu_{ij} \neq \mu_{i+1,j}$ holds. Note that a flow shop is a special case of a job shop, whereas an open shop is not.

Example 2.3. Consider a flow shop with two jobs and two machines. For the first job, the processing requirements of its two operations are $p_{11} = 2$ and $p_{21} = 1$ and for the second job they are $p_{11} = 2$ and $p_{21} = 2$. Figure 2.8(a) gives the best schedule that can be obtained when scheduling job 1 before job 2. We see that the last job is ready at time unit 6. If, however, we schedule the jobs the other way around, i.e., if job 2 is scheduled before job 1, then we obtain a schedule in which the last job finishes at time unit 5. Hence, if it is our goal to minimize the total time required for handling all jobs, the latter schedule outperforms the first one. □

Job characteristics. The job characteristics are specified by $\beta = \beta_1\beta_2\beta_3\beta_4\beta_5$.

- $\beta_1 \in \{\circ, pmtn\}$. The abbreviation *pmtn* stands for preemption. Preemption is allowed if we have $\beta_1 = pmtn$, and it is not allowed if we have $\beta_1 = \circ$. If preemption is allowed, then the processing of operations may be interrupted, i.e., jobs may be processed in multiple stages.

- $\beta_2 \in \{\circ, prec\}$. Variable β_2 indicates whether ($\beta_2 = prec$) or not ($\beta_2 = \circ$) precedence relations exist between the jobs. Precedence relations are generally specified by an acyclic directed graph. The jobs that have to be processed form the node set of the graph, and an arc from job i to job j specifies that job i has to be completed before we can start processing job j.

- $\beta_3 \in \{\circ, r_j\}$. If we have $\beta_3 = r_j$, then jobs have release dates. The release date of a job is the time from which it is available for processing. If, on the other hand, we have $\beta_3 = \circ$, then no release dates are specified, which is equivalent to assuming $r_j = 0$ for all $j \in J$.

- $\beta_4 \in \{\circ, s_{ij}\}$. If we have $\beta_4 = s_{ij}$, then jobs have sequence-dependent setup times. This means that if job i is scheduled directly before job j on a given machine, then it takes the machine s_{ij} time units before it can start processing job j after finishing job i. Furthermore, the setup time required before processing the first job j on a machine is given by s_{0j}, and the clean-up time after finishing the last job j on a machine is given by s_{j0}. We do not have sequence-dependent setup times if $\beta_4 = \circ$.

- $\beta_5 \in \{\circ, p_j = 1, p_{ij} = 1\}$. If we have $\alpha_1 \in \{\circ, P, Q\}$, then β_5 can be given by $p_j = 1$, which means that each job has unit processing requirement. Analogously, if we have $\alpha_1 \in \{O, F, J\}$, then β_5 can be given by $p_{ij} = 1$, which means that each operation has unit processing requirement. The processing requirements may be arbitrary if $\beta_5 = \circ$.

In the literature many other job characteristics are considered and added to β. We have restricted ourselves to giving the ones that are most common.

Optimality criterion. Before explaining optimality criterion γ, we introduce some definitions. If a job j consists of only one operation, then its completion time in a schedule is denoted by C_j. If it consists of multiple operations, then the completion time of operation o_{ij} is given by C_{ij} and the completion time of the job is given by $C_j = \max_i C_{ij}$. Knowing the completion time of a job j, we can compute its lateness $L_j = C_j - d_j$. The lateness is negative if a job is finished before its due date. If this is undesirable, we consider the tardiness $T_j = \max(L_j, 0)$ of job j. The earliness $E_j = \max(d_j - C_j, 0)$ of job j expresses how much time the job is finished before its due date. Finally, if we are only interested in whether or not a job is on time, we have U_j, which is zero if $C_j \leq d_j$ and one, otherwise.

Based on these definitions, we can now specify the optimality criterion by $\gamma \in \{\max_j \gamma_j, \max_j w_j \gamma_j, \sum_j \gamma_j, \sum_j w_j \gamma_j\}$ with $\gamma_j \in \{C_j, L_j, T_j, E_j, U_j\}$ and with w_j indicating that the weighted version of the problem is considered, where the weights are non-negative integers. The optimality criterion has to be minimized. For instance, if we have $\gamma = \max_j w_j C_j$, then we are given a non-negative weight for each job and the problem is to minimize the maximum weighted completion time. We

note that we usually write 'γ_{max}' instead of '$max_j \gamma_j$' and that C_{max} is called the makespan.

A well-studied and basic machine scheduling problem is MULTIPROCESSOR SCHEDULING. In this problem we have to schedule n independent jobs on m identical, parallel machines, such that the completion time of the last job is minimal. Hence, the problem can be written as $P||C_{max}$. Another classical problem is JOB SHOP SCHEDULING, which can be written as $J||C_{max}$.

Finally, consider $1|s_{ij}|C_{max}$, in which we are asked to sequence jobs with setup times on a single machine such that the makespan is minimized. This problem is equivalent to TSP. To see this, consider the following mapping between the two problems, bearing in mind that in a problem instance of TSP C denotes the set of cities and d_{ij} the distance from city i to j. Let the set of jobs be given by $C - \{1\}$, let $s_{ij} = d_{ij}$ for all jobs i, j, and let the processing requirements of the jobs be arbitrary. Furthermore, take $s_{0i} = d_{1i}$ and $s_{i0} = d_{i1}$ for each job i. Then the makespan of a schedule is equal to the total processing requirement plus the length of the tour that starts at city 1 and visits the other $n - 1$ cities in the order in which the corresponding jobs are processed on the single processor.

Insertion, move, and swap. For many machine scheduling problems, such as $1|s_{ij}|C_{max}$, it holds that in an optimal schedule a machine starts processing at time zero and it does not contain idle time between two successive jobs (operations), where setup times are not counted as idle time. In that case a solution can be represented by m sequences of jobs that give the order in which the jobs are processed on the machines. For $P||C_{max}$ an even simpler representation suffices, as the order in which the jobs are processed on the machines is irrelevant. We are only interested in a partition of the jobs that specifies which jobs are processed by the same machine. As an optimal schedule for JOB SHOP SCHEDULING may contain idle time, it is less trivial to apply local search to this problem. In Chapter 4 we discuss how to tackle JOB SHOP SCHEDULING with local search.

If a solution is defined by m sequences of jobs, then we can use the insertion as well as the swap neighborhood function. In the insertion neighborhood function, neighbors are generated by deleting a job from a sequence and inserting it either at another position in the same sequence or in the sequence of another machine. The swap neighborhood function generates neighbors by swapping two arbitrary jobs that may or may not be in the same sequence. Both neighborhood functions can also be used if a solution is defined by a partition of the jobs in m subsets. In that case the insertion neighborhood function is called the move neighborhood function. As the swap neighborhood function does not change the number of jobs that are assigned to a machine, it has the unfavorable behavior that if one starts with a solution in which no jobs are assigned to a machine, the machine remains unused. Therefore, the swap neighborhood of a solution is often extended with its insertion (move) neighborhood. Note that the insertion neighbors of a solution of $1|s_{ij}|C_{max}$ are the same as the node-insertion neighbors of the corresponding TSP problem instance, except that city n is not selected for a node insertion.

Consider a machine scheduling problem, for instance $P||C_{max}$, in which we are asked to schedule multiple machines and in which it is our objective to minimize the makespan. When, in choosing the cost function, we simply stick to the makespan of the schedule, we get a poor performance of iterative improvement, as a schedule in which multiple machines have the same makespan is locally optimal even though other machines may still be empty. To overcome this problem, we can resort to the following (slightly) different cost function. Let a critical machine be defined as a machine that does not finish before C_{max}. Then, we redefine the cost function, such that it returns two values, namely the makespan of the schedule and the number of critical machines. A schedule S has lower cost than schedule S' if the cost of S is lexicographically smaller than the cost of S', where we take the makespan as the first objective and the number of critical machines as the second objective. In Section 2.4 we present another example in which it is effective to slightly adapt the cost function.

2.3 Steiner Tree Problem in Graphs

For TSP we presented neighborhood functions that are based on the obvious solution space containing all tours of n cities and we used the cost function that simply assigns to a tour its length. For MULTIPROCESSOR SCHEDULING we indicated that it may be useful to resort to a non-obvious cost function. It is often also useful to switch to a non-obvious solution space. Consider, for instance, the modified version of TSP in which it need no longer be possible to travel from each city i to each other city j. Finding a starting solution then becomes more complicated. In general, it is even NP-hard because it corresponds to finding a Hamiltonian cycle in a graph. Furthermore, the k-change neighborhood size can be much smaller than for standard TSP, which has a negative effect on the performance of local search algorithms. A possible remedy to these problems is to enlarge the solution space with infeasible solutions. More specifically, we can add the infeasible tours containing one or more non-existing arcs. However, we have to be careful that the better solutions are feasible. This can be realized by penalizing each non-existing arc in a tour, i.e., by defining the cost of a tour as the total distance of the existing arcs plus the number of non-existing arcs times a given large value.

In this section we define a neighborhood function for STEINER TREE PROBLEM IN GRAPHS (STG). This illustrates another general approach in which we switch to a non-obvious solution space. The common idea behind the approach is the use of an indirect representation of a solution. It may be the case, such as for STG, that once we know some features of a solution it is easy to construct the best solution with these features. In that case, we can use these features as an indirect representation of a solution and we can reduce the solution space by switching from the set of all solutions to the set of all relevant feature values.

Definition 2.4 [STEINER TREE PROBLEM IN GRAPHS (STG)]. Given is an edge-weighted graph $G = (V, E)$ with a weight $w(e) \in \mathbb{N}$ for each $e \in E$. The node set V is partitioned into two sets R and U. Set R contains the special nodes and U the Steiner

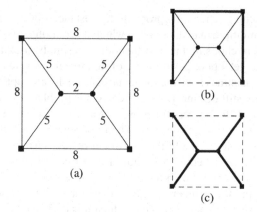

Figure 2.9. (a) Graph with four special nodes (boxes) and two Steiner nodes (circles). (b) Minimum spanning tree on the special nodes. (c) Optimal Steiner tree.

nodes. A Steiner tree of G is a subtree $T = (V_T, E_T)$ of G that includes all special nodes, i.e., $R \subseteq V_T \subseteq V$. The problem is to find a Steiner tree $T = (V_T, E_T)$ of G with minimum weight $w(T) = \sum_{e \in E_T} w(e)$. □

Example 2.4. Consider the graph given in Figure 2.9. The corners of the square are the four special nodes, and the two nodes inside the square are the Steiner nodes. The weight of an edge is given by the Euclidean distance between the two nodes it connects, where the square is assumed to have sides of length eight.

At first sight, it may seem that the optimal Steiner tree is given by the minimum spanning tree on the set R of special nodes. However, the weight of this tree is 24, whereas the minimum spanning tree on all nodes is only 22, which is smaller. □

Insertion. Several polynomial-time algorithms exist for deriving a spanning tree of minimum weight for a given edge-weighted graph $G = (V, E)$. A popular algorithm for deriving such a tree, called a minimum spanning tree, works as follows. It starts with the tree T that contains a single arbitrary node from V. Next, it recursively adds to T a shortest edge of which one node is already contained in T and one node is not; see Figure 2.10. The algorithm can be implemented such that it computes a minimum spanning tree in $\mathcal{O}(|V|^2)$ time.

The problem formulation of STG suggests that we define the solution space S as the set of all Steiner trees in G and define the cost function as w. However, we can use the set of Steiner nodes in a Steiner tree as an indirect solution representation. Once we are given a set $A \subseteq U$ of Steiner nodes, we can use the minimum spanning tree algorithm of Figure 2.10 on the subgraph of G induced by $A \cup R$ to derive an optimal Steiner tree MST(A) containing exactly these Steiner nodes and all special nodes.

Now we can apply local search to the solution space 2^U containing all subsets $A \subseteq U$ of Steiner nodes and the cost function that defines the cost of a set $A \in 2^U$

algorithm Minimum Spanning Tree

begin
 $V_T := \{v\}$ for some $v \in V$;
 $E_T := \emptyset$;
 while $V_T \neq V$ **do**
 begin
 determine $\{v',v\} \in V_T \times V$ that minimizes $w(\{v',v\})$;
 $V_T := V_T \cup \{v\}$;
 $E_T := E_T \cup \{v',v\}$;
 end
end;

Figure 2.10. An algorithm for deriving a minimum spanning tree $T = (V_T, E_T)$ of a graph $G = (V, E)$. The algorithm is used for STG to derive a solution from an indirect representation.

as the weight of the Steiner tree MST(A). Obviously, an optimal solution to this problem gives an optimal solution to the original problem. This is important as it means that we are still tackling the same problem. For a neighborhood function we can use the insertion neighborhood function which is defined as follows. The neighbors of a set A are the sets obtained from A by adding or removing a Steiner node, i.e., by inserting a Steiner node in either A or $U \setminus A$. Figure 2.11 gives the iterative improvement algorithm for the insertion neighborhood function.

2.4 Graph Coloring

In 1852, the student Francis Guthrie informed his brother Frederick that he had been considering the problem of coloring the countries of a map in such a way that no two neighboring countries have the same color and such that the number of colors used is minimal. He pointed out that four colors seemed to be sufficient to do the job, and he wondered if this was generally true. Despite its apparent simplicity, it took mathematicians more than a century to settle this appealing problem. Finally, after many failed attempts, Appel and Haken proved in 1976 that this conjecture is indeed true. However, this does not end the fascinating quest for a proof of the 'four color theorem'. As the proof of Appel and Haken makes use of a computer, it is hidden in a black box and difficult to verify by humans. For some sceptics, this means it is not fully acceptable as a watertight proof.

The map coloring problem described can be modeled as a graph problem in the following way. For a given map M, we define a graph G such that every region of M corresponds to a node in G and the graph contains an edge $\{i, j\}$ if and only if i and j are adjacent regions in M. It can be verified that the resulting graph G is planar. Now the map coloring problem corresponds to coloring the nodes of G with

algorithm Iterative Improvement for STG

begin
 initialize $A \subseteq U$;
 $T_A := \mathrm{MST}(A)$;
 repeat
 generate an $A' \in N(A)$;
 $T_{A'} := \mathrm{MST}(A')$;
 if $w(T_{A'}) < w(T_A)$ **then** $A := A'$;
 until $w(T_{A'}) \geq w(T_A)$ for all $A' \in N(A)$;
 end;

Figure 2.11. Iterative improvement algorithm for STG with the insertion neighborhood function $N(A) = \{A \cup \{v\} \mid v \in U \setminus A\} \cup \{A \setminus \{v\} \mid v \in A\}$. MST$(A)$ denotes the minimum spanning tree derived by the algorithm of Figure 2.10 for the subgraph of G that is induced by $A \cup R$.

a minimum number of colors, such that no two nodes that are connected by an edge have the same color. The latter problem is called MINIMUM GRAPH COLORING. It can also be verified that for an arbitrary planar graph MINIMUM GRAPH COLORING can be modeled as a map coloring problem. This implies that the two problems are equivalent. In its general form, MINIMUM GRAPH COLORING is formalized as follows.

Definition 2.5 [MINIMUM GRAPH COLORING (MGC)]. Given is a graph $G = (V, E)$. Find a coloring of the nodes, such that no two adjacent nodes have the same color and the number of colors used is minimal, i.e., find a partition V_1, V_2, \ldots, V_k of V into a minimum number k of independent sets. □

For planar graphs we are well capable of finding high-quality solutions. We can decide in linear time whether a coloring using one or two colors exists and, if this is the case, construct one. Furthermore, it is also possible to construct a five-coloring in linear time. According to the four color theorem, such a coloring cannot be optimal. Nevertheless, a quadratic-time algorithm exists that constructs a coloring using only four colors. The question whether or not this coloring can be improved to three is NP-complete.

For general graphs, MGC is much harder to approximate. It has been shown that it is very unlikely that a polynomial-time algorithm exists with a performance ratio better than $|V|^{1-\varepsilon}$ for any constant $\varepsilon > 0$. Note that the naive algorithm that assigns a different color to each node has a performance ratio of exactly $|V|$.

In 1879, Kempe published the first 'proof' of the four color conjecture. Although eleven years later Heawood showed that Kempe's proof was wrong, the proof introduces Kempe chains, which are still frequently used in the context of MGC. Before we explain a Kempe chain, consider the colored graph given in Fig-

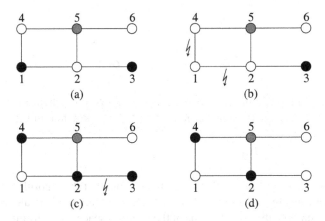

Figure 2.12. (a) Feasible coloring of a graph with the colors white, black, and gray. (b) Infeasible coloring obtained by changing the color of node 1 from black to white. (c,d) Transformation of infeasible coloring (b) to a feasible coloring.

ure 2.12(a) and suppose that we recolor node 1 from black to white. We then get an infeasible coloring as both end points in edges $\{1,2\}$ and $\{1,4\}$ are colored white; see Figure 2.12(b). To repair this, we assign nodes 2 and 4 the original color of node 1, which is black. Again, this results in an improper coloring as the end points of edge $\{2,3\}$ are both black. As before, we undo this violation by assigning node 3 the original color of node 2, which is white. We now end up with the feasible coloring depicted in Figure 2.12(d).

The algorithm presented for constructing a feasible coloring after recoloring one of the nodes can be formalized as follows. Let G be a properly colored graph, and suppose that we want to recolor node v from color c to color c' with $c \neq c'$. Furthermore, let G' be the graph obtained from G by removing all nodes that do not have the color c or c'. The component in subgraph G' containing v is called a Kempe chain. For the coloring given in Figure 2.12(a), Figure 2.13 depicts this chain for the case that we want to color node 1 white. To construct a new coloring we perform a Kempe chain interchange. This means that we swap the colors in the chain: nodes with the color c are colored c' and vice versa. It can be verified that the resulting coloring is again feasible. Note that the Kempe chain interchange on the Kempe chain of Figure 2.13 indeed results in the feasible coloring given in Figure 2.12(d).

The discussion above gives rise to the neighborhood function in which a coloring C' is a neighbor of coloring C if C' can be obtained from C by performing a Kempe chain interchange. Furthermore, to construct a starting solution, we can use the following greedy algorithm. Let π be an arbitrary ordering of the nodes of the graph, and let the colors with which we want to color the graph be numbered from one onwards. Then the greedy algorithm colors the nodes in the order specified by π, such that each node is assigned the minimum color that is not already assigned to any of its neighbors. Obviously, the algorithm uses at most $\Delta + 1$ colors, where Δ is

Figure 2.13. Kempe chain on which we perform a Kempe chain interchange in case we want to change the color of node 1 from black to white in the coloring of Figure 2.12(a).

the maximum degree of any node in the graph. Note that the algorithm can derive different colorings by varying π. This is a welcome feature of the algorithm as it enables us to explore different regions of the solution space by performing multiple runs of iterative improvement. We return to this in Chapter 7.

We now arrive at the following implementation of iterative improvement or any other local search algorithm. We use the greedy algorithm for constructing an initial solution using at most $\Delta + 1$ colors, we let the solution space contain all feasible colorings using at most $\Delta + 1$ colors, and we apply the Kempe chain neighborhood function presented. What remains is to choose a cost function. A straightforward option is to define the cost of a coloring as the number of colors used. However, this is not a good choice. Contrary to the exponential size solution space, the cost function has only a very small range since it can take at most $\Delta + 1$ values, namely $1, 2, \ldots, \Delta + 1$. This gives an indication that the solution space will contain large plateaus, which will have a negative effect on the performance of local search algorithms. The fact that we have large plateaus can also be seen by realizing that we cannot improve the cost of a coloring by a small number of moves in the case that each color is used several times and in different parts of the graph.

To circumvent this problem we can define the cost of a coloring as a $\Delta + 1$ tuple $(c_1, c_2, \ldots, c_{\Delta+1})$, where c_i gives the number of nodes with color i. Furthermore, we consider the cost $f(C)$ of a coloring C to be smaller than the cost $f(C')$ of a coloring C' if it is lexicographically smaller. Hence, the cost function f, which clearly has a much larger range than the original cost function, stimulates the use of colors with a high index number. Note that if a locally optimal solution uses k colors, then it uses each of the largest k colors and none of the first $\Delta - k + 1$ colors. Obviously, a globally optimal coloring for this cost function uses a minimum number of colors, which means that it is also an optimal coloring for the original cost function. Hence, by changing the cost function we still solve the same problem, i.e., MGC.

2.5 Uniform Graph Partitioning

The effectiveness of a local search algorithm strongly depends on the risk of getting stuck in a poor local optimum. For many problems, this risk can be reduced by applying an approach called variable-depth search. In this section we illustrate this approach for UNIFORM GRAPH PARTITIONING.

Definition 2.6 [UNIFORM GRAPH PARTITIONING (UGP)]. Given is an edge-weighted graph $G = (V, E)$ with $|V| = 2n$ and a weight $w(e) \in \mathbb{N}$ for each edge $e \in E$. Find a partition of V into two subsets V_1 and V_2 with $|V_1| = |V_2| = n$, such that the sum of the weights of the edges that have one endpoint in V_1 and one endpoint in V_2 is minimal, i.e., such that

$$c(V_1, V_2) = \sum_{e \in E \cap (V_1 \times V_2)} w(e)$$

is minimal. □

This minimization problem is equivalent to the maximization variant of the problem, as can be seen as follows. Let $G = (V, E)$ be a weighted graph with maximum weight $W = \max_{e \in E} w(e)$, and let K be the complete graph on the same node set V, in which the weight of an edge e is given by $W - w(e)$ if $e \in E$ and by W otherwise. The cost of a partition (V_1, V_2) in one variant of the problem is then equivalent to $n^2 W$ minus the cost of this partition in the other variant. This implies that both problems are indeed equivalent.

An obvious neighborhood function for UGP is the swap neighborhood function in which the neighbors of a partition (V_1, V_2) are obtained by interchanging a node from V_1 with a node from V_2. To obtain more powerful neighborhood functions, one can generalize the swap neighborhood function to the k-swap neighborhood function in a similar way as we generalized 2-change to k-change for TSP. This means that for any given $k \geq 1$ the k-swap neighbors of a partition (V_1, V_2) are obtained by interchanging at most k nodes from V_1 with just as many nodes from V_2. The size of the k-swap neighborhood is given by $\sum_{i=1}^{k} \binom{n}{i}^2 = \Theta(n^{2k})$, where $\binom{n}{i}^2$ gives the number of alternatives for interchanging i nodes in V_1 with i nodes in V_2. It follows that while the quality of the local optima of the k-swap neighborhood function increases with k, this is at the expense of a considerable increase in the neighborhood size and, consequently, also in the running time of iterative improvement.

Variable-depth search gives an alternative, effective approach for making the swap neighborhood function more powerful. The main idea is that we allow multiple swaps to be performed simultaneously. We call the resulting neighborhood function Kernighan-Lin, after its two inventors. For each i with $1 \leq i \leq n$, the Kernighan-Lin neighborhood of a partition (V_1, V_2) contains exactly one partition that is obtained from (V_1, V_2) by exchanging i nodes from V_1 with i nodes from V_2. This is a drastic reduction in comparison with the k-swap neighborhood, which contains for each i with $1 \leq i \leq k$ all $\binom{n}{i}^2$ partitions that can be obtained by exchanging i nodes from V_1 with i nodes from V_2.

The n Kernighan-Lin neighbors of a partition (V_1, V_2), written as $(V_1^{(i)}, V_2^{(i)})$ for $1 \leq i \leq n$, can be determined in n steps as follows. In Step i we derive partition $(V_1^{(i)}, V_2^{(i)})$ by performing the best possible swap on (V_1, V_2) if $i = 1$ and on the partition $(V_1^{(i-1)}, V_2^{(i-1)})$ derived in the previous step if $2 \leq i \leq n$. We are not allowed to select a node that has already been selected before for a swap. By the best possible swap we mean the swap of the nodes from $V_1^{(i-1)}$ and $V_2^{(i-1)}$ that produces the

maximum decrease in cost or, if such a swap does not exist, a minimum increase in cost. In the case of a tie, a tie-breaking rule is used to choose a unique pair of nodes that is swapped. The neighborhood consists of n partitions because, after n swaps, all nodes of the graph have been selected for an exchange and, by definition, each node may only be selected once. In other words, the neighbors are obtained by performing iterative improvement using the best-improvement pivoting rule and the swap neighborhood function. However, unlike the standard iterative improvement algorithm, we have the additional constraint that each node may only be selected once for a swap and that we accept a best cost-increasing neighbor if we are in a local optimum.

Consider iterative improvement with the Kernighan-Lin neighborhood function. An iteration of the algorithm can be formalized as follows. Let $g_{V_1,V_2}(v_1,v_2)$ be the change in cost resulting from swapping nodes $v_1 \in V_1$ and $v_2 \in V_2$ in a given partition (V_1,V_2). Then the first neighbor of a partition (V_1,V_2), which we denoted by $(V_1^{(1)},V_2^{(1)})$, is obtained by swapping the nodes $v_1^{(1)}$ and $v_2^{(1)}$, where $v_1^{(1)}$ and $v_2^{(1)}$ minimize $g_{V_1,V_2}(v_1^{(1)},v_2^{(1)})$, and the ith neighbor $(V_1^{(i)},V_2^{(i)})$ with $2 \le i \le n$ is obtained by swapping the nodes $v_1^{(1)},v_1^{(2)},\ldots,v_1^{(i)}$ and $v_2^{(1)},v_2^{(2)},\ldots,v_2^{(i)}$, where $v_1^{(i)}$ and $v_2^{(i)}$ minimize $g_{V_1^{(i-1)},V_2^{(i-1)}}(v_1^{(i)},v_2^{(i)})$ over all $v_1^{(i)} \in V_1^{(i-1)} \cap V_1$ and $v_2^{(i)} \in V_2^{(i-1)} \cap V_2$. Now the difference in cost between partition (V_1,V_2) and the ith neighbor can be defined as

$$G(i) = \sum_{j=1}^{i} g_{V_1^{(j-1)},V_2^{(j-1)}}(v_1^{(j)},v_2^{(j)}).$$

Hence, we find that iterative improvement stops in the considered iteration if $G(i) \ge 0$ for all $1 \le i \le n$. Otherwise, it swaps the nodes $v_1^{(1)},v_1^{(2)},\ldots,v_1^{(i)}$ and $v_2^{(1)},v_2^{(2)},\ldots,v_2^{(i)}$, where i minimizes $G(i)$ if the best-improvement pivoting rule is used and where i is the first value for which $G(i) < 0$ if the first-improvement pivoting rule is used.

Since the Kernighan-Lin neighborhood of a solution has only size n and since each neighbor can be determined in $\mathcal{O}(n^2)$ time, the Kernighan-Lin neighborhood function is much more manageable than the k-swap neighborhood function for $k \ge 4$. Furthermore, like k-swap, the Kernighan-Lin neighborhood function is more powerful than the swap neighborhood function. To substantiate this, observe that a partition (V_1,V_2) is swap optimal if and only if the first neighbor in its Kernighan-Lin neighborhood has higher cost. In addition, we observe that, unlike the neighborhood functions we discussed in the previous sections, the Kernighan-Lin neighborhood function need not be symmetric.

Example 2.5. Consider Figure 2.14(a), which depicts a weighted graph $G = (V,E)$ consisting of two components $C_1 = \{1,2,3,4,5,6\}$ and $C_2 = \{7,8,9,10,11,12\}$ of six nodes each and a partition (V_1,V_2) with cost $c(V_1,V_2) = 4$. Obviously, this partition is not optimal as the two components C_1 and C_2 define a partition with cost zero. However, it is locally optimal with respect to the swap neighborhood function:

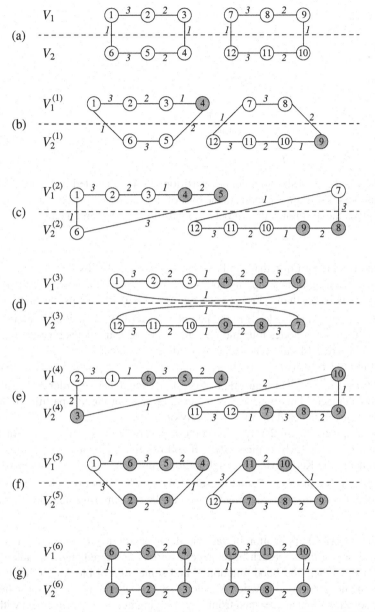

Figure 2.14. (a) Partition of the nodes of a graph into the sets $V_1 = \{1,2,3,7,8,9\}$ and $V_2 = \{4,5,6,10,11,12\}$. The numbers along the edges define their weights. (b)-(g) Partitions in the Kernighan-Lin neighborhood of (V_1, V_2).

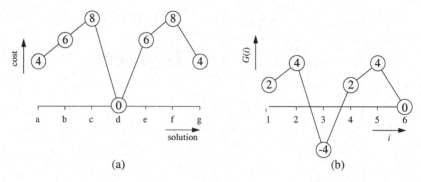

Figure 2.15. (a) The six neighbors of partition (V_1, V_2) depicted in Figure 2.14(a). The numbers inside a node indicate the cost of the solutions. (b) The function $G(i)$, which gives the difference in cost between partition (V_1, V_2) and its ith neighbor.

the best possible partition that can be obtained from (V_1, V_2) by performing a swap is the one resulting from swapping nodes four and nine and the cost of this partition is 6, which is larger than $c(V_1, V_2) = 4$. Figure 2.14(b) shows this partition, which is the first partition $(V_1^{(1)}, V_2^{(1)})$ in the neighborhood of (V_1, V_2). When defining the other neighbors, the nodes four and nine are no longer candidates for performing a swap. In Figure 2.14 this is indicated by shading these nodes.

In Figure 2.14(c)-(g) we also define the other $n - 1 = 5$ neighbors. Note that the last partition $(V_1^{(6)}, V_2^{(6)})$ is equivalent to (V_1, V_2), except that the order of the subsets has been changed. This is generally the case because in the last partition of V all nodes have been selected for a swap exactly once.

The graph in Figure 2.15(a) shows the $n = 6$ neighbors of (V_1, V_2) with their cost, and Figure 2.15(b) depicts $G(i)$. It follows that if iterative improvement is applied using the Kernighan-Lin neighborhood function, then the optimal partition $(V_1^{(3)}, V_2^{(3)})$ given in Figure 2.14(d) is selected as the neighbor of partition (V_1, V_2). This holds for the first-improvement as well as the best-improvement pivoting rule.

□

With the described application of variable-depth search to UGP in mind, it is now easy to describe the approach in its general form. As ingredients, variable-depth search needs a neighborhood function N and some kind of blocking rule. It then defines the neighborhood function N', such that $N'(s)$ contains exactly those neighbors of solution s that are reached from s by applying iterative improvement with the best-improvement pivoting rule, where the iterative improvement is only allowed to make moves that are not prohibited by the blocking rule and where a best non-improving, admissible move is chosen in the case that all admissible moves lead to neighbors that are worse than the current solution.

2.6 *Stable Configuration in Hopfield Networks

In this section we show how the problem of finding a locally optimal solution arises in neural computing. A Hopfield network is an example of a neural network. An interesting type of Hopfield network can be specified by an edge-weighted simple graph $G = (V, E)$. The weight $w(e)$ of an edge $e \in E$ may be either a positive or a negative integer. The same holds for the threshold value T_v that is assigned to each node $v \in V$. A configuration f assigns to each node v the value $f(v) = -1$ or $f(v) = 1$. In the former case, node v is considered to be in the off-state, while in the latter case it is in the on-state. We call node v stable in a configuration f if we have

$$f(v) = 1 \Rightarrow \sum_{\{u,v\}\in E} w(\{u,v\})f(u) \geq T_v$$

and

$$f(v) = -1 \Rightarrow \sum_{\{u,v\}\in E} w(\{u,v\})f(u) \leq T_v.$$

If all nodes are stable in a given configuration f, then f is said to be stable.

An interesting question is whether a stable configuration always exists and, if this is the case, how to find one. To answer these questions, consider the cost function that assigns to a configuration f the cost

$$\sum_{\{u,v\}\in E} w(\{u,v\})f(u)f(v) - \sum_{v\in V} T_v f(v). \tag{2.2}$$

Next, suppose that we change the state of node v. Then calculus yields that the resulting change in cost is given by

$$\Delta_v = (1 - 2f(v))(\sum_{\{u,v\}\in E} w(\{u,v\})f(u) - T_v).$$

We note that in $\sum_{\{u,v\}\in E} w(\{u,v\})f(u)$ variable v is fixed, which is not the case in the sum $\sum_{\{u,v\}\in E} w(\{u,v\})f(u)f(v)$ in (2.2). Using the definition of Δ_v, we can derive that changing the state of an unstable node v results in a cost increase. To see this, first consider the case where $f(v) = -1$. We then have $1 - 2f(v) = 3$ and, as v is unstable, also $\sum_{\{u,v\}\in E} w(\{u,v\})f(u) > T_v$. This clearly implies $\Delta_v > 0$. Similarly, if $f(v) = 1$ holds for unstable node v, then we have $1 - 2f(v) = -1$ and $\sum_{\{u,v\}\in E} w(\{u,v\})f(u) < T_v$, which again yields $\Delta_v > 0$. Hence, flipping the state of an unstable node does indeed result in an increase in cost. By similar arguments, it can be shown that flipping the state of a stable node does not improve the cost of a configuration, i.e., a cost increase can only be obtained by flipping the state of an unstable node. As a result, we obtain that a configuration is stable if and only if it is a locally optimal solution with respect to maximizing the cost function given by (2.2) and the neighborhood function that flips the state of a node. This means that we can find a stable configuration for the type of Hopfield network we consider by applying iterative improvement. This answers the questions postulated above as to whether stable configurations exist and how to find one. Note that it follows from our discussion that this application of iterative improvement corresponds to

repeatedly flipping the state of unstable nodes in the Hopfield network, which is the algorithm that is traditionally applied in Hopfield networks.

2.7 Bibliographical Notes

The first use of a local search algorithm for tackling a combinatorial optimization problem dates back to the 1950s. At that time, 2-Opt was introduced by Flood [1956] and Croes [1958] and 3-Opt by Bock [1958]. The real effectiveness of k-Opt was shown by Lin [1965]. Lin applies k-Opt to several problem instances for $k = 2, 3$, and 4. The experiments show that 3-Opt yields a good trade-off between running time and solution quality. For 2-Opt the shorter running time is at the expense of a considerable loss in quality, while for 4-Opt the increase in quality is only minor compared to the increase in running time. Nicholson [1971] extend the concept of exchange strategies to a more general class of permutation problems, including network layout design, scheduling, vehicle routing, and cutting stock problems. The Or neighborhood function for TSP proposed in Section 2.1 was introduced by Or [1976]. The book published by Lawler, Lenstra, Rinnooy Kan & Shmoys [1985] can be viewed as the classical textbook on TSP. Reinelt [1994] published a book emphasizing the many computational approaches to the problem, and Johnson & McGeoch [1997] wrote a chapter on the application of local search to TSP. For a more recent book on TSP, we refer to Gutin & Punnen [2002].

In Section 2.2, we presented a three-field classification $\alpha|\beta|\gamma$ for specifying machine scheduling problems. This notation is introduced by Graham, Lawler, Lenstra & Rinnooy Kan [1979]. For a more detailed discussion of the different alternatives we have for the machine environment, job characteristics, and optimality criterion, we refer to Pinedo [1995] and Brucker [1995].

The insertion neighborhood function we discussed for STG is proposed by Osborne & Gillett [1991]. Furthermore, the algorithm we presented for computing a minimum spanning tree is due to Prim [1957] and Dijkstra [1959]. Papadimitriou & Steiglitz [1982] present a nice discussion of the problem of finding a minimum spanning tree.

Regarding the computational complexity of STG, Garey, Graham & Johnson [1977] prove that even the Euclidean variant is strongly NP-hard. In this variant, called EUCLIDEAN STEINER TREE, the nodes represent points in a plane and the weight of an edge $\{i, j\}$ is defined as the (discretized) Euclidean distance between points i and j. Furthermore, Bern & Plassmann [1989] prove that STG is APX-complete, which implies that STG \in APX $-$ PTAS provided that P \neq NP. Robins & Zelikovsky [2000] derive an approximation algorithm with a performance ratio of $1 + \frac{\ln 3}{2}$. For EUCLIDEAN STEINER TREE, on the other hand, Arora [1998] proves that it does admit a PTAS, which implies that, unless P $=$ NP, this strongly NP-hard problem belongs to the class PTAS $-$ FPTAS; see Figure 2.16.

In Section 2.4 we started by giving a short history of MGC. A more elaborate history is given by Saaty & Kainen [1977]. Furthermore, we refer to Gibbons [1985] for a more detailed discussion of the indicated equivalence between the map coloring problem and MINIMUM GRAPH COLORING ON PLANAR GRAPHS. We note that

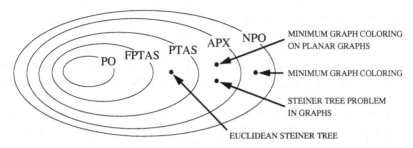

Figure 2.16. Position of some problems in the existing approximability classes in the case that P \neq NP. The approximability classes are discussed in Appendix B.

this equivalence was already noticed by Kempe [1879].

Appel and Haken proved the famous four color theorem; see Appel & Haken [1977], Appel, Haken & Koch [1977], and Appel & Haken [1989]. Furthermore, an algorithm for constructing a coloring for a planar graph using four colors is presented by Robertson, Sanders, Seymour & Thomas [1997]. The running time of this algorithm is quadratic in $|V|$. As it is easy to derive a linear-time algorithm for constructing a coloring using only one or two colors if such a coloring exists, the latter result implies that the problem of coloring planar graphs allows a polynomial-time $\frac{4}{3}$-approximation algorithm. However, since an approximation algorithm with a better performance ratio would be able to distinguish in polynomial time 3-colorable graphs from 4-colorable graphs and since Garey, Johnson & Stockmeyer [1976] prove that deciding whether a planar graph is 3-colorable is NP-complete, this is the best we can hope for. Hence, if P \neq NP, then the problem is in the approximability class APX $-$ PTAS. This is visualized in Figure 2.16. For planar graphs, we also note that a linear-time algorithm for coloring the graph with five colors is given by Chiba, Nishizeki & Saito [1981].

Next, consider MGC for general graphs. Halldórsson [1993] proves the existence of a polynomial-time $\mathcal{O}(|V|\frac{(\log\log|V|)^2}{(\log|V|)^3})$-approximation algorithm. Furthermore, Bellare, Goldreich & Sudan [1998] prove that, unless P = NP, no polynomial-time algorithm exists with a performance bound of $|V|^{1/7}$. As a result, MGC is in NPO $-$ APX provided that P \neq NP. Feige & Kilian [1998] strengthen the non-approximability result. They prove that no polynomial-time algorithm exists with a performance bound of $|V|^{1-\varepsilon}$ for any constant $\varepsilon > 0$. However, this result holds under the condition that ZPP \neq NP instead of the weaker condition P \neq NP, where ZPP (Zero-error Probabilistic Polynomial time) is the class of languages for which a randomized algorithm exists that decides in polynomial expected time whether a given string is in the language and is guaranteed to give the correct answer. For more details on the class ZPP, we refer to Papadimitriou [1994].

We conclude our discussion of MGC by restating that Kempe chains are introduced in the 'proof' of the four color theorem by Kempe [1879] and that an error in this proof was discovered by Heawood [1890].

In Section 2.5 we discuss UGP. Garey, Johnson & Stockmeyer [1976] prove that the problem we obtain when dropping the constraint $|V_1| = |V_2| = n$ in the maximization variant of UGP is strongly NP-hard. It can easily be shown that this result remains valid if we add the constraint $|V_1| = |V_2| = n$. Since UGP is equivalent to its maximization problem, as we have indicated in Section 2.5, this yields that UGP is also strongly NP-hard.

As indicated by its name, the Kernighan-Lin neighborhood function presented for UGP has been introduced by Kernighan & Lin [1970]. Although many different problems have been tackled by neighborhood functions based on variable-depth search, we would like to draw attention to the paper of Lin & Kernighan [1973]. They present a local search algorithm for TSP, later to be known as the Lin-Kernighan algorithm, that uses the approach and for which Johnson & McGeoch [1997] report that it was recognized as the world champion heuristic algorithm until about 1990.

In Section 2.6 we discuss a restricted type of Hopfield network. In its general form, the weighted graph defining a Hopfield network may be directed and it may contain loops. Hopfield networks have been introduced by Hopfield [1982], who also proved the result of Section 2.6, i.e., that finding a stable configuration corresponds to finding a local optimum with respect to the cost function given by (2.2) and the neighborhood function that flips the state of nodes. For more details on Hopfield networks we refer to Wasserman [1989].

2.8 Exercises

1. Consider MULTIPROCESSOR SCHEDULING, which can be written as $P||C_{max}$.

 a) How large is the solution space, i.e., how many partitions exist of n jobs over m machines?

 b) How large is the move neighborhood size of a solution?

 c) Is the move neighborhood function exact? Substantiate your answer by means of a proof or a counterexample.

2. Show for TSP that the neighborhood size of node insertion is $n(n-3)$ for any $n \geq 4$.

3. Consider the following problem.

 Definition 2.7 [KNAPSACK PROBLEM]. Given are a set A of items, a bound $B \in \mathbb{N}^+$, and for each item $i \in A$ a size $s_i \in \mathbb{N}^+$ and a value $v_i \in \mathbb{N}^+$. A subset $A' \subseteq A$ is said to be feasible if its total size is at most B, i.e., if $\sum_{i \in A'} s_i \leq B$. Find a feasible subset $A' \subseteq A$ with maximum value $\sum_{i \in A'} v_i$. □

 If we want to tackle this problem by local search, then we can use the swap neighborhood function in which a subset A'' is a neighbor of a subset A' if and only if it can be obtained by replacing an item in A' with an item that is not in A'. Alternatively, we can use the move neighborhood function in which a subset A''

Figure 2.17. (a) Problem instance of EUCLIDEAN TSP with 16 cities. (b)-(e) Four tours for this problem instance.

is a neighbor of a subset A' if and only if it can be obtained either by removing an item from A' or by adding an item to it. Both neighborhood functions are defined on the solution space containing all feasible subsets.

a) Give a major drawback of the swap neighborhood function.

Consider the problem instance with $A = \{1, 2, 3\}$, $B = 3$, $(s_1, s_2, s_3) = (2, 1, 2)$, and $(v_1, v_2, v_3) = (1, 2, 3)$.

b) Give the move neighborhood graph and move transition graph for this problem instance.

c) Give the diameter of the neighborhood graph and the potential of the transition graph.

d) Let a start solution be chosen uniformly at random. Determine for each possible solution the probability that it occurs as the final solution when iterative improvement is applied with the best-improvement pivoting rule and the move neighborhood function.

4. Consider the problem instance of EUCLIDEAN TSP depicted in Figure 2.17(a). The problem instance consists of 16 cities, and the distance between two neighboring cities is one, regardless of whether they are on the same horizontal line or on the same vertical line. Determine for the four tours depicted in Figures 2.17 whether they are locally optimal with respect to the 2-change neighborhood function and with respect to the node-insertion neighborhood function.

5. Consider the following practical problem. Given are n data blocks, which need to be fetched from m multi-zone hard disks. Each data block i with $1 \le i \le n$ is stored on all m hard disks. The time required for reading block i from disk j with $1 \le j \le m$ is given by p_{ij}. This time need not be constant due to the multi-zone character of the hard disks. If a disk reads blocks i and i' successively, then an overhead of s time units is required to move the reading head from the end of block i to the beginning of block i'. Now we have to determine which data block to read from which disk, such that the time that all disks are finished is as early as possible. For this it is allowed that a data block is split into several parts, where each part is read from a different disk (notice that in this case for each part a switch time s is required).

Write this problem in the machine scheduling notation presented in Section 2.2.

6. Figure 2.10 gives an algorithm for deriving a minimum spanning tree of a graph $G = (V,E)$ in $\mathcal{O}(|V|^2)$ time. Prove that this time complexity is asymptotically optimal when expressed in $|V|$ (and not, for example, in $|E|$). In other words, prove that any algorithm for deriving a minimum spanning tree has a time complexity of $\Omega(|V|^2)$.

7. Consider UGP. In Section 2.5 we claimed that each Kernighan-Lin neighbor of a partition (V_1, V_2) can be derived in $\mathcal{O}(n^2)$ time. As in the same section we indicated that the size of the swap neighborhood is $\Theta(n^2)$, this claim is true if we can determine the change in cost resulting from swapping any pair of nodes in constant time.

 a) Prove that we can determine the cost effect of swapping any two nodes in constant time if we keep for each node $v \in V_i$, $i \in \{1,2\}$, its external cost

 $$E(v) = \sum_{\{v,v'\} \in E \cap (V_1 \times V_2)} w(\{v,v'\})$$

 and its internal cost

 $$I(v) = \sum_{\{v,v'\} \in E \cap (V_i \times V_i)} w(\{v,v'\}).$$

 b) Show that after performing a swap we can update the external and internal cost value of any node in constant time.

8. Design two variable-depth search neighborhood functions for MULTIPROCESSOR SCHEDULING, one based on the move neighborhood function and one based on the swap neighborhood function.

9. A regular graph is a graph in which all nodes have the same degree. Hence, like the term isotropic, which we defined in Definition 2.3, the term regular indicates that a graph is well structured. Prove that an isotropic graph is always regular and that the reverse need not be true.

3

Indirect Solution Representations

We can distinguish two types of neighborhood functions: those that are defined on representations that directly define a solution and those that are defined on representations that indirectly define a solution. Except for the insertion neighborhood function proposed for STG, all examples given in Chapter 2 are based on direct representations.

Indirect solution representations are typically used when we can identify a feature for all the solutions, such that for a given value of this feature a best corresponding solution can be derived in polynomial time. To make this more precise, let f be such a feature and let equivalence relation \sim_f on the solution space be defined by $s \sim_f s'$ if and only if solutions s and s' have the same value for feature f. The equivalence classes induced by \sim_f partition the solution space into disjoint subsets. This gives rise to basing local search on the new solution space containing all possible values of feature f, where a value of f represents a best solution in the corresponding equivalence class. By the definition of f, the solution represented by some value of f can be derived in polynomial time.

To illustrate the use of indirect solution representations, we discussed STG in Section 2.3. For this problem we can define an equivalence relation \sim on the set of Steiner trees by $T \sim T'$ if and only if Steiner trees T and T' contain the same Steiner nodes. We can now use a subset A of the Steiner nodes as an indirect representation of a best Steiner tree T_A in the corresponding equivalence class, i.e., of a best Steiner tree T_A with the property that it contains all Steiner nodes in A and no other Steiner nodes. We indicated that the tree T_A can be computed in polynomial time for a

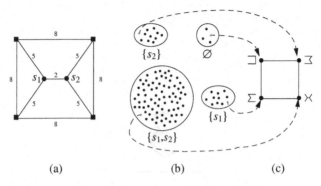

(a) (b) (c)

Figure 3.1. (a) Graph with four special nodes (boxes) and two Steiner nodes, s_1 and s_2. (b) Four equivalence classes implied by equivalence relation \sim, where $T \sim T'$ if and only if Steiner trees T and T' contain the same Steiner nodes. A subset below a class indicates the Steiner nodes that are contained in the Steiner trees of that class. (c) Neighborhood graph implied by the insertion neighborhood function and the solution space $2^{\{s_1,s_2\}} = \{\emptyset, \{s_1\}, \{s_2\}, \{s_1,s_2\}\}$. A solution represents a best solution in the corresponding equivalence class.

given subset A as it corresponds to deriving a minimum spanning tree. As a possible neighborhood function, we proposed the insertion neighborhood function in Section 2.3. For the graph of Figure 3.1(a), which is the graph depicted in Figure 2.9, Figure 3.1(b) gives the equivalence classes induced by \sim. We have one equivalence class with 4 solutions, two with 11 solutions, and one with 75 solutions. Hence, for this toy example the solution space reduces from 101 solutions to only 4 solutions when we use the indirect representation implied by \sim. The neighborhood graph for this example when using this indirect representation and the insertion neighborhood function is depicted in Figure 3.1(c).

In this chapter we give examples showing the use of an indirect representation for three machine scheduling problems, namely $R||\sum_j w_j C_j$, $1||\sum_j E_j + T_j$, and $J||C_{\max}$. All indirect representations are based on the approach described above, in which each indirect solution representation corresponds to a best solution in an equivalence class.

3.1 Sum of Weighted Completion Time on Unrelated Machines

The first machine scheduling problem we address is $R||\sum_j w_j C_j$. For this problem we only have to consider schedules that do not contain idle time between jobs and in which each machine starts processing at time zero. Hence, as a direct solution representation we can use m sequences of jobs, where sequence i specifies which jobs are to be processed by machine i and in what order. However, the solution space can be reduced considerably by resorting to an indirect solution representation. The key to this indirect representation is the observation that an optimal order to process a

set of jobs on a single machine, say machine i, can be computed in $\mathcal{O}(n\log n)$ time
by putting the jobs in non-decreasing order of the value of the ratio p_{ij}/w_j. This
order is called the shortest weighted processing time (SWPT) order. The result is
proved in Theorem 3.1. We can now base a local search algorithm on the indirect
solution space containing all possible assignments of jobs to machines and on the
cost function that gives the cost $\sum_j w_j C_j$ of an optimal schedule corresponding to
an assignment, which can be computed in $\mathcal{O}(n\log n)$ time. Possible neighborhood
functions are the move and swap neighborhood functions. Note that the equiva-
lence relation on which the indirect representation is based relates direct solution
representations that have the same assignment of the jobs to the machines.

Theorem 3.1. *An optimal schedule for* $1||\sum_j w_j C_j$ *is obtained by sequencing the
jobs in SWPT order.*

Proof. We prove the theorem by contradiction. Hence, assume that we have an
optimal one-machine schedule S in which the jobs are not sequenced in SWPT or-
der. Since the jobs are not in SWPT order, two jobs j and j' exist, such that j' is
the direct successor of job j in the schedule and $p_j/w_j > p_{j'}/w_{j'}$. Let S' be the
schedule obtained by swapping these jobs. Swapping the jobs only affects the com-
pletion times of jobs j and j'. The completion time of job j increases by $p_{j'}$ and
the completion time of job j' decreases by p_j. As a result, the cost of the schedule
increases by $w_j p_{j'} - w_{j'} p_j$. As $p_j/w_j > p_{j'}/w_{j'}$ implies $w_j p_{j'} < w_{j'} p_j$, this cost
increase is smaller than zero. This implies that schedule S' has a lower cost than S,
contradicting our assumption that S is optimal. □

3.2 *Symmetric Earliness and Tardiness on a Single Machine

For the machine scheduling problem discussed above the main challenge was to
assign the jobs to the machines. Given the set of jobs that have to be processed by
a machine i, the problem of deriving an optimal schedule for machine i was easy
as it corresponds to sequencing the jobs in SWPT order with no idle time. We now
consider $1||\sum_j E_j + T_j$ in which each job has a due date d_j and in which we are
asked to schedule the jobs on a single machine, such that the sum of earliness and
tardiness is minimal. Without loss of generality, we may assume that $p_j \leq d_j$ holds
for each job j, as a problem instance with $p_j > d_j$ is equivalent to the same problem
instance with the due date of j replaced by p_j and with a constant penalty of $p_j - d_j$
added to the cost function. For problem $1||\sum_j E_j + T_j$, determining an assignment
of the jobs to the machines is no longer an issue, as we only have one machine.
However, to determine a schedule for a single machine it no longer suffices to find
an ordering of the jobs. We also have to specify a starting time for each job as
it may be favorable to add idle time between two successive jobs. We present a
polynomial-time algorithm for determining an optimal schedule for a given order in
which the jobs have to be processed. This implies that we can tackle $1||\sum_j E_j + T_j$
by performing local search on the solution space containing all permutations, i.e.,
orderings of the jobs. Obviously, the insertion and swap neighborhood function are
both candidates for use in the local search algorithm.

Consider an arbitrary order in which the n jobs have to be processed. Without loss of generality, we assume that the jobs are numbered, such that job j is the jth job in this ordering. We present an algorithm that assigns a completion time C_j to each job j, such that the corresponding schedule is optimal. Note that deriving a completion time C_j for job j is equivalent to deriving its starting time σ_j because $C_j = \sigma_j + p_j$ as preemptions are not allowed. We first introduce some terminology. A schedule consists of one or more blocks, which is a maximal sequence of jobs that are scheduled contiguously. This means that the first job of a block is either the first job in the schedule or it is preceded by idle time and, similarly, the last job of a block is either the last job in the schedule or it is succeeded by idle time. The scheduling algorithm schedules the jobs one at a time in increasing order. We define S_i as the schedule obtained after scheduling job i.

Let B_1, B_2, \ldots, B_t be the blocks of a schedule S_i. We partition the jobs of a block B_l into the two subsets $Late(l)$ and $Early(l)$. The former subset contains the late jobs and the latter the early jobs. Formally, these two subsets are given by $Late(l) = \{ j \in B_l | C_j > d_j \}$ and $Early(l) = \{ j \in B_l | C_j \le d_j \}$. Hence, the jobs from $Late(l)$ benefit from advancing block B_l completely and the jobs from $Early(l)$ benefit from postponing the block. For example, the schedule of Figure 3.2(d) consists of three blocks. The first block B_1 consists of job $1 \in Early(1)$ and jobs $2, 3 \in Late(1)$, block B_2 contains job $4 \in Early(2)$ and job $5 \in Late(2)$, and block B_3 contains only job $6 \in Early(3)$. For a block B_l, we define $N_{early}(l)$ and $N_{late}(l)$ as the number of jobs in $Early(l)$ and $Late(l)$, respectively. Furthermore, $f(S)$ denotes the cost $\sum_j E_j + T_j$ of schedule S.

We now present the scheduling algorithm, which we call Block Shifting. In the first iteration, the algorithm schedules the first job, such that it meets its due date, i.e., $C_1 = d_1$; see Figure 3.2(a). In iteration i with $i \ge 2$ the algorithm constructs schedule S_i from S_{i-1} as follows. First of all, consider the case $C_{i-1} + p_i \le d_i$. In this case it is possible to let job i meet its due date without this affecting the scheduling of the first $i - 1$ jobs as defined by S_{i-1}. We make use of this possibility, i.e., we schedule job i such that $C_i = d_i$. Note that if $C_{i-1} + p_i < d_i$, then schedule S_i contains idle time between job $i - 1$ and i. Furthermore, we have $f(S_i) = f(S_{i-1})$. An essential property of the algorithm will be that it satisfies the following invariant.

Invariant. *For an arbitrary block B_l we have that it either starts at time zero or satisfies $N_{late}(l) < N_{early}(l)$.*

Note that if block B_l starts at time zero, then it is the first block in the schedule. Furthermore, observe that the invariant implies that a schedule derived by the algorithm is locally optimal with respect to advancing one of the blocks. This is true because advancing a block B_l for one time unit yields a cost change of $N_{late}(l) - N_{early}(l)$. As a consequence of the proposed assignment of job i to S_{i-1}, we find that $N_{early}(t)$ is increased by one if job i is added to the last block B_t of S_{i-1} and that a new block B_{t+1} is initiated with $N_{late}(t+1) = 0$ and $N_{early}(t+1) = 1$ otherwise. In both cases the invariant is not violated.

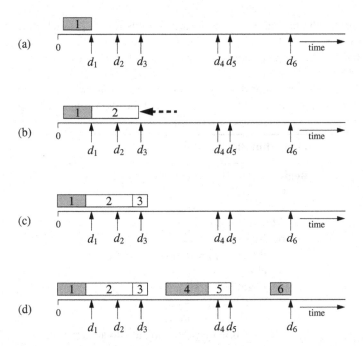

Figure 3.2. (a)-(c) First three iterations of the Block-Shifting algorithm. (d) Final schedule. A job in block B_l is in $Early(l)$ if the corresponding box is shaded and it is in $Late(l)$ otherwise.

Next, assume that $C_{i-1} + p_i > d_i$. Then we schedule job i directly after job $i-1$. This implies that $C_i = C_{i-1} + p_i$ and that job i is added to the last block B_t of schedule S_{i-1}. Furthermore, we have that $i \in Late(t)$, which implies that $N_{late}(t)$ is increased by one. As indicated in Figure 3.2(b), this may cause a violation of the invariant. We now show how to repair it. Assume that the invariant is violated, i.e., B_t does not start at time zero and $N_{late}(t) \geq N_{early}(t)$. Because the invariant holds for schedule S_{i-1}, we have that S_{i-1} satisfies $N_{late}(t) < N_{early}(t)$. Hence, we can strengthen the assumption $N_{late}(t) \geq N_{early}(t)$ to $N_{late}(t) = N_{early}(t)$. This implies that advancing the entire block B_t does not change the cost of the schedule. We advance the block, without changing the cost of the schedule, until one of the following three conditions are satisfied.

- Block B_t starts at time zero.
- A job $j \in Late(t)$ meets its due date.
- Block B_{t-1} and B_t are merged, i.e., the completion time of the last job of block B_{t-1} is equal to the starting time of the first job of block B_t.

The invariant clearly holds if the first case occurs. In the second case $N_{late}(t)$ decreases by one and $N_{early}(t)$ increases by one, which implies $N_{late}(t) < N_{early}(t)$.

Next, consider the last case. As the invariant holds for S_{i-1}, block B_{t-1} starts at time zero or $N_{late}(t-1) < N_{early}(t-1)$. Combined with $N_{late}(t) = N_{early}(t)$ this means that for the merged block B'_{t-1} we also have that it starts at time zero or $N_{late}(t-1) < N_{early}(t-1)$.

The Block-Shifting algorithm is given in Figure 3.3. Furthermore, Figure 3.2 visualizes for an example the first three iterations of the algorithm and its final schedule.

algorithm Block Shifting

begin
 $C_1 := d_1$;
 $i := 2$;
 while $i \leq n$ **do**
 begin
 if $C_{i-1} + p_i \leq d_i$ **then** $C_i := d_i$;
 else
 begin
 $C_i := C_{i-1} + p_i$;
 $B_t := $ last block;
 while invariant violated **do** advance B_t;
 end
 $i := i+1$;
 end
end;

Figure 3.3. The Block-Shifting algorithm, which is used for $1||\sum_j E_j + T_j$ to derive a solution from an indirect representation. Without loss of generality, it is assumed that the jobs have to be scheduled in increasing order

Theorem 3.2. *The schedule computed by the Block-Shifting algorithm returns a schedule that minimizes $\sum_j E_j + T_j$ for a given ordering of the jobs.*

Proof. We prove by induction on i that S_i is an optimal schedule for the first i jobs. The induction hypothesis clearly holds for $i = 1$. We now show that it also holds for any $i \geq 2$ by deriving a contradiction from the assumption that an optimal schedule S'_i of the first i jobs has strictly lower cost than S_i. We distinguish the same two cases as in the presentation of the algorithm.

Case 1: $C_{i-1} + p_i \leq d_i$. We then have $f(S_i) = f(S_{i-1})$, where, by the induction hypothesis, schedule S_{i-1} is optimal for the first $i-1$ jobs. Obviously, removing job i from S'_i does not increase the cost of schedule S'_i. Hence $f(S'_i) \geq f(S_{i-1}) = f(S_i)$, which contradicts our assumption that S'_i is better than S_i.

Case 2: $C_{i-1} + p_i > d_i$. The algorithm constructs S_i by first constructing from S_{i-1} schedule R in which job i is scheduled directly after job $i-1$. Next, the last block

B_t of R is advanced if the invariant is violated. As stated above, this operation does not change the cost of the schedule. Hence, $f(R) = f(S_i)$ holds. We derive a contradiction by showing that $f(S_i') \geq f(R)$.

Assume that the completion time C_i' of job i in schedule S_i' is at least the completion time C_i of this job in R, i.e., $C_i' \geq C_i$. Let S_{i-1}' be defined as schedule S_i' minus job i. Since $C_i' \geq C_i > d_i$, we have $f(S_i') = f(S_{i-1}') + C_i' - d_i$. As $f(S_{i-1}') \geq f(S_{i-1})$, by the induction hypothesis, and as $C_i' \geq C_i$, we get $f(S_i') \geq f(S_{i-1}) + C_i - d_i = f(R)$.

Assume, on the other hand, that $C_i' < C_i$. This implies that $C_j' < C_j$ holds for each job j in B_t, where C_j' and C_j give the completion times of job j in S_i' and R, respectively. Note that this implies that block B_t does not start at time zero in schedule R. We first prove by contradiction the auxiliary result that the jobs from B_t also form (a part of) a block in S_i'. Assume that jobs j and $j+1$ are the first two jobs from B_t that are not direct successors in schedule S_i'. We show that the gap between these jobs can be removed, such that the cost of the schedule decreases. This contradicts our assumption that S_i' is optimal, which proves our auxiliary result that the jobs from B_t indeed form (a part of) a block in optimal schedule S_i'.

Let $V(B_t, j)$ be defined as the set of jobs from B_t that are smaller than or equal to j, i.e., that have to be processed before job j. The invariant implies for B_t that $N_{late}(t) < N_{early}(t)$ holds at each stage of its construction, where $N_{late}(t)$ and $N_{early}(t)$ refer to B_t at the stage under consideration. Because advancing a block may add jobs to the set $Early(t)$ but does not remove jobs from this set, this implies that the set $V(B_t, j)$ contains more jobs from $Early(t)$ than from $Late(t)$. Furthermore, as long as the completion time of job j in S_i' is strictly smaller than its completion time in R, none of the jobs in $V(B_t, j) \cap Early(t)$ meets its due date in S_i'. This means that, as a result of postponing the jobs from $V(B_t, j)$ in schedule S_i' until the completion time of job j coincides with the starting time of job $j+1$ (which is still smaller than its completion time in R because $C_i' < C_i$), we obtain that the jobs from $Early(t)$ cause a reduction of the cost that is strictly larger than the increase that can be caused by postponing the jobs from $Late(t)$. Hence, S_i' is not an optimal schedule, which yields a contradiction. This proves our claim that the jobs from B_t also form (a part of) a block in S_i'.

This result, the assumption $C_i' < C_i$, and the property that $N_{late}(t) \leq N_{early}(t)$ holds in R imply that the total earliness and tardiness penalty incurred by the jobs from B_t is not smaller in S_i' than in R. Furthermore, by the induction hypothesis, we have $f(S_l') \geq f(S_l)$, where l is i minus the number of jobs in B_t and S_l' is obtained from S_i' by removing the jobs in B_t. Because S_l is contained in R, these two observations yield $f(S_i') \geq f(R)$, which was to be proved. \square

3.3 Job Shop Scheduling

JOB SHOP SCHEDULING, which can be written as $J||C_{\max}$, is one of the more difficult combinatorial optimization problems. Although the problem has already been introduced in Section 2.2, we repeat its definition here.

Definition 3.1 [JOB SHOP SCHEDULING]. Given are m machines and a set of n jobs $J = \{1, 2, \ldots, n\}$, where each job $j \in J$ consists of a set $\{o_{1j}, o_{2j}, \ldots, o_{m_j j}\}$ of m_j operations. Furthermore, for each operation o_{ij} we are given a machine μ_{ij} on which it has to be processed and a processing requirement p_{ij}, where we have $\mu_{ij} \neq \mu_{i+1,j}$ for all $j \in J$ and $1 \leq i < m_j$. The problem is to find a schedule with minimum makespan. This means that we have to find a starting time σ_{ij} for each operation o_{ij}, such that for all $j \in J$ and $1 \leq i < m_j$ we have

$$\sigma_{ij} + p_{ij} \leq \sigma_{i+1,j},$$

such that for all $j, j' \in J$, $1 \leq i \leq m_j$, and $1 \leq i' \leq m_{j'}$ with $o_{ij} \neq o_{i'j'}$ we have

$$(\mu_{ij} = \mu_{i'j'}) \wedge (\sigma_{ij} \leq \sigma_{i'j'}) \Rightarrow \sigma_{ij} + p_{ij} \leq \sigma_{i'j'},$$

and such that

$$\max_{j \in J} (\sigma_{m_j j} + p_{m_j, j})$$

is minimal. □

The best known local search algorithms are based on a representation of a problem instance as a node-weighted disjunctive graph $G = (V, A, E)$ in the following way.

- The node set V contains a source node s, a target node t, and a node for each operation o_{ij} with $j \in J$ and $1 \leq i \leq m_j$. The weights of s and t are zero and the weights of the other nodes are given by the processing requirements of the corresponding operations. The nodes s and t are two fictitious operations with zero processing requirement that serve as pointers to the beginning and end of the schedule.

- The arc set A contains an arc (o, o') if $o = s$ and o' is the first operation of a job, if o and o' are two successive operations of the same job, and if o is the last operation of a job and $o' = t$.

- The edge set E contains an edge between any two operations o_{ij} and $o_{i'j'}$, $o_{ij} \neq o_{i'j'}$, that have to be processed on the same machine, i.e., for which $\mu_{ij} = \mu_{i'j'}$ holds.

The interpretation of an arc $(o, o') \in A$ is that operation o has to be processed before operation o' because of the precedence constraints imposed by the jobs, and the interpretation of an edge $\{o, o'\} \in E$ is that operation o must be processed before operation o' or vice versa because a machine can only process one job at a time.

Example 3.1. Figure 3.4 gives the node-weighted disjunctive graph corresponding to a problem instance with two jobs and three machines. Both jobs consist of three operations, and all operations have a processing requirement of one time unit. Furthermore, operations o_{11} and o_{32} have to be processed by machine 1, operations o_{21} and o_{22} by machine 2, and operations o_{31} and o_{12} by machine 3; see Table 3.1. □

Let, for a given schedule, $\pi_i(l)$ define the lth operation processed by machine i. Then we call the tuple $\Pi = (\pi_1, \pi_2, \ldots, \pi_m)$ the machine ordering of the schedule.

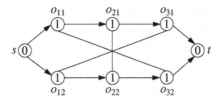

Figure 3.4. Node-weighted disjunctive graph corresponding to the problem instance of Example 3.1.

We present a polynomial-time algorithm for finding an optimal schedule for a given machine ordering Π. Note that such a schedule may contain idle time. Using the algorithm, we can, as for problem $1||\sum_j E_j + T_j$ in Section 3.2, base a local search algorithm on the solution space that contains all machine orderings. However, unlike for $1||\sum_j E_j + T_j$, we also use this algorithm to restrict our solution space to a subset of all machine orderings and to formulate the neighborhood function. A machine ordering Π can be modeled in the disjunctive graph corresponding to the problem instance considered by orienting the edges of E. If π_i states that operation o is processed before operation o' on machine i, then the edge $\{o, o'\} \in E$ is replaced by the arc (o, o'). We define E_Π as the set of arcs obtained by orienting the edges from E according to Π. We denote the resulting digraph by $D_\Pi = (V, A \cup E_\Pi)$. To illustrate this, consider Example 3.1. Figure 3.5 gives the corresponding digraph for each possible machine ordering. For instance, the first digraph corresponds to the machine ordering defined by $\pi_1 = (o_{11}, o_{32})$, $\pi_2 = (o_{21}, o_{22})$, and $\pi_3 = (o_{12}, o_{31})$.

Obviously, if a feasible schedule exists that satisfies Π, then D_Π is acyclic and the cost C_{\max} of the schedule is at least the length of a longest directed path from s to t in D_Π, where the length of a path is defined as the total weight of the nodes the path visits. Conversely, if D_Π is acyclic, then we can derive a feasible schedule for which the cost is given by the length of a longest directed path from s to t. This schedule can be constructed by defining the completion time of an operation o as the length of a longest path from s to o. Figure 3.6 gives the schedules that are obtained in this way for the acyclic digraphs in Figure 3.5. Summarizing, we have that machine ordering Π implies a feasible schedule if and only if D_Π is acyclic. Furthermore, if D_Π is acyclic, then the problem of deriving an optimal schedule for Π corresponds to deriving a longest path from s to any node $o \in V$ in digraph D_Π. This problem can be solved in $\mathcal{O}(|A \cup E_\Pi|)$ time by the following algorithm, which, in addition to using the property that D_Π is acyclic, also uses the property that all nodes in D_Π are reachable from s.

Initially, we give node s the label zero. All other nodes are left unlabeled. Next, we repeat the following procedure. Find an unlabeled node o for which all predecessors in the digraph are labeled. If such a node cannot be found, then terminate. Otherwise, assign to node o the label $w(o) + \max_{(o', o) \in A \cup E_\Pi} label(o')$, where $w(o)$ is the weight of node o, i.e., the processing requirement of operation o, and $label(o')$ is the label of node o'.

Table 3.1. Problem instance discussed in Example 3.1.

J	$\{1,2\}$
M	$\{1,2,3\}$
job 1	$\{o_{11}, o_{21}, o_{31}\}$
job 2	$\{o_{12}, o_{22}, o_{32}\}$

p_{ij}	1
$(\mu_{11}, \mu_{21}, \mu_{31})$	$(1,2,3)$
$(\mu_{12}, \mu_{22}, \mu_{32})$	$(3,2,1)$

The algorithm presented, which is given in pseudo-code in Figure 3.7, assigns a label to each node o, and this label gives the length of a longest path from s to o, i.e., the completion time of operation o in an optimal schedule for Π. A longest path in a graph is also called a critical path. Correspondingly, we say that an arc is critical if it occurs in some critical path.

We now define the following local search algorithm for a problem instance of JOB SHOP SCHEDULING. As the solution space, we take the set of machine orderings $\Pi = (\pi_1, \pi_2, \ldots, \pi_m)$ for which a feasible schedule exists, i.e., for which the corresponding digraph D_Π is acyclic. For Example 3.1 this means that the solution space only contains the machine orderings corresponding to the first four digraphs in Figure 3.5. The cost of solution Π is defined as the length of a longest path in the digraph D_Π, which can be computed in $\mathcal{O}(|A \cup E_\Pi|)$ time. Finally, we define the neighborhood function as the swap neighborhood function presented for machine scheduling problems in Section 2.2, except that we only swap two operations o and o' in solution Π if they are both immediate successors on some machine and if the arc (o,o') is on a critical path in D_Π. By the definition of JOB SHOP SCHEDULING, two successive operations of the same job have to be processed by different machines. So we have $(o,o') \in E_\Pi$ and not $(o,o') \in A$. Furthermore, if $(o,o') \in E_\Pi$ is on a critical path for some pair o,o' of operations, then o and o' are immediate successors on a machine. Hence, Π' is a neighbor of Π if $D_{\Pi'}$ can be obtained from D_Π by reversing an arc $(o,o') \in E_\Pi$ on a critical path in D_Π. Figure 3.8 gives the neighborhood graph for the problem instance of Example 3.1.

A motivation for this neighborhood function, which we call critical-path swap, is that reversing a non-critical arc in D_Π cannot lead to a better solution Π' as the digraph $D_{\Pi'}$ still contains each critical path from D_Π. Furthermore, the following theorem states that reversing a critical arc from E_Π always yields a feasible solution. This does not have to be the case if a non-critical arc from E_Π is reversed, as cyclic digraph 7 in Figure 3.5 can, for instance, be obtained from acyclic digraph 1 by reversing the arc (o_{11}, o_{32}).

Theorem 3.3. *Let D_Π be an acyclic digraph and let $D_{\Pi'}$ be the digraph obtained by reversing a critical arc $e = (o,o') \in E_\Pi$. Then $D_{\Pi'}$ is also acyclic.*

Proof. We prove the theorem by contradiction. Hence, assume that $D_{\Pi'}$ is cyclic. As, by definition, two successive operations of a job are not assigned to the same machine, we have that $(o,o') \notin A$ and thus that (o,o') is not an arc in $D_{\Pi'}$. Furthermore, each cycle in $D_{\Pi'}$ has to contain the arc (o',o) because any other cycle would

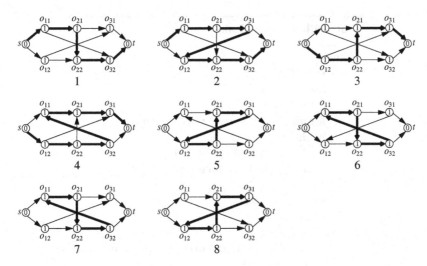

Figure 3.5. All digraphs for the problem instance of Example 3.1. In each digraph we indicate its longest path or, if the digraph is cyclic, its cycle.

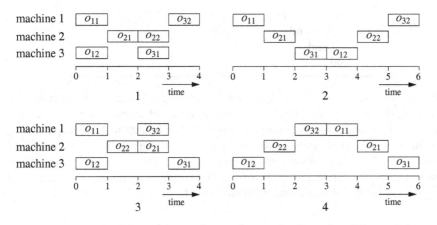

Figure 3.6. Schedules corresponding to the acyclic digraphs of Figure 3.5.

also be in D_Π, which would contradict the assumption that D_Π is acyclic. These observations imply that $D_{\Pi'}$ contains a path $p = (o, x_1, x_2, \ldots, x_l, o')$ with $l \geq 1$ and $x_i \notin \{o, o'\}$. However, p can also be found in D_Π, and the length of p is strictly larger than the length of arc $e = (o, o')$. This contradicts the assumption that e is on a critical path of D_Π. □

3.4 Bibliographical Notes

Theorem 3.1 has been proved by Smith [1956]. Lenstra, Rinnooy Kan & Brucker [1977] prove that $P2||\sum_j w_j C_j$, which is a special case of problem $R||\sum_j w_j C_j$ dis-

algorithm Longest Path

begin
 initialize N_{pred};
 $label(s) := 0$;
 $C := \emptyset$;
 for all $o' \in \{o' \mid (s,o') \in A \cup E_\Pi\}$ **do**
 begin
 $N_{pred}(o') := N_{pred}(o') - 1$;
 if $N_{pred}(o') = 0$ **then** $C := C \cup \{o'\}$;
 end
 while $C \neq \emptyset$ **do**
 begin
 select $o \in C$;
 $label(o) := w(o) + \max_{(o',o) \in A \cup E_\Pi} label(o')$;
 $C := C \setminus \{o\}$;
 for all $o' \in \{o' \mid (o,o') \in A \cup E_\Pi\}$ **do**
 begin
 $N_{pred}(o') := N_{pred}(o') - 1$;
 if $N_{pred}(o') = 0$ **then** $C := C \cup \{o'\}$;
 end
 end
end;

Figure 3.7. An $\mathcal{O}(|A \cup E_\Pi|)$ algorithm for deriving a longest path in an acyclic digraph $D_\Pi = (V, A \cup E_\Pi)$ from s to any other node, where each node is reachable from s. Note that s has only outgoing arcs. In the algorithm, $N_{pred}(o')$ gives the number of unlabeled predecessors of a node o' and C gives the set of unlabeled nodes for which all predecessors are labeled. The algorithm is used for JOB SHOP SCHEDULING to derive a solution from an indirect representation, i.e., to derive an optimal schedule for a given machine ordering Π.

cussed in Section 3.1, is NP-hard in the ordinary sense. Garey & Johnson [1979] mention that if the number of machines is part of the input, then this special case becomes NP-hard in the strong sense.

The algorithm proposed for solving $1||\sum_j E_j + T_j$ for a given permutation of the jobs is due to Garey, Tarjan & Wilfong [1988]. They also prove Exercise 3(b) and that the problem $1||\sum_j E_j + T_j$ is NP-hard. Based on similar ideas as those used by Garey, Tarjan & Wilfong [1988], Davis & Kanet [1993] show that the problem $1||\sum_j w_j E_j + w'_j T_j$ can be solved efficiently for a given permutation of the jobs, as well.

The disjunctive graph formulation discussed in Section 3.3 is given by Roy & Sussmann [1964]. Theorem 3.3 has been proved by Van Laarhoven [1988] and Van

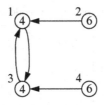

Figure 3.8. Neighborhood graph for the problem instance discussed in Example 3.1. The number above a node gives the number of the corresponding digraph in Figure 3.5. The number inside the node gives the cost of the solution, i.e., the length of a longest path in the digraph. It follows that the two schedules shown on the left in Figure 3.6 are optimal. Obviously, the neighborhood graph is asymmetric and not strongly connected.

Laarhoven, Aarts & Lenstra [1992]. Garey, Johnson & Sethi [1976] prove that JOB SHOP SCHEDULING is already strongly NP-hard for two machines. Gonzalez & Sahni [1978] show that this remains true if preemption would be allowed.

3.5 Exercises

1. Consider $R||T_{max}$. We say that jobs are scheduled in earliest due date (EDD) order on a single machine if they are scheduled in non-decreasing order of their due date.

 a) Prove that an optimal schedule for $1||T_{max}$ is obtained by sequencing jobs in EED order.

 b) Design a neighborhood function for $R||T_{max}$ that is based on an indirect representation.

2. Consider the following problem. Given are n uniform parallel machines, a bound B, and a set of weighted jobs. Find a schedule that maximizes the total weight of the jobs that have been processed at time B, where preemption is allowed.

 a) Prove that $Q|pmtn|C_{max}$ is polynomially solvable.

 b) Design a neighborhood function for the problem described that is based on an indirect representation.

3. Consider $1||\sum_j E_j + T_j$. In Section 3.2 we proved Theorem 3.2, which states that for a given ordering of the jobs the Block-Shifting algorithm computes an optimal schedule. Based on this result, we can tackle $1||\sum_j E_j + T_j$ by performing local search on the solution space containing all permutations of the jobs.

 a) Show that applying Block Shifting to the permutation in which the jobs are ordered by increasing due date does not guarantee optimality.

 b) Prove that if all jobs have the same processing requirement, then applying

Table 3.2. Problem instance considered in Exercise 4.

J	$\{1,2,3\}$		p_{i1}, p_{i3}	1
M	$\{1,2,3\}$		p_{i2}	2
job 1	$\{o_{11}, o_{21}, o_{31}\}$		$(\mu_{11}, \mu_{21}, \mu_{31})$	$(1,2,3)$
job 2	$\{o_{12}, o_{22}\}$		(μ_{12}, μ_{22})	$(2,1)$
job 3	$\{o_{13}, o_{23}, o_{33}\}$		$(\mu_{12}, \mu_{22}, \mu_{32})$	$(2,3,1)$

Figure 3.9. Schedule for the problem instance defined by Table 3.2.

Block Shifting to the permutation in which the jobs are ordered by increasing due date does guarantee optimality.

4. Table 3.2 defines a problem instance of JOB SHOP SCHEDULING with three jobs and three machines.

 a) Give the node-weighted disjunctive graph G that models the problem instance.

 b) Give the machine ordering Π of the schedule depicted in Figure 3.9.

 c) Give the digraph D_Π that is obtained from G if we model Π.

 d) Is Π locally optimal with respect to the critical-path swap neighborhood function? Motivate your answer.

4

Properties of Neighborhood Functions

In this chapter we analyze several neighborhood functions. More specifically, we prove properties on the diameter and the connectivity of neighborhood graphs and we relate the neighborhood graphs of different neighborhood functions. In Section 1.1 we have seen that these properties can be considered as important necessary conditions for the effectiveness of neighborhood functions. In addition, these properties enhance our understanding of the neighborhood functions and they can also add value to the results proved later in this book. For instance, the $\mathcal{O}(n)$ diameter we prove in Section 4.1 for the k-change neighborhood graph makes the exponential number of iterations that iterative improvement may require when using the k-change neighborhood function even more intriguing. For the latter result we refer to Chapter 6.

We first focus in Section 4.1 on neighborhood functions for TSP and next in Section 4.2 on the neighborhood function that we presented in Section 3.3 for job shop scheduling.

4.1 Traveling Salesman Problem

In this section we derive properties of the node-insertion and k-change neighborhood function for TSP. First consider the former neighborhood function. In Exercise 2 of Chapter 2 we asked the reader to prove the following result.

Theorem 4.1. *For* TSP, *each node in the node-insertion neighborhood graph has degree* $n(n-3)$ *for any* $n \geq 4$. □

Furthermore, the following theorem states that the diameter is $n-2$. Note that a finite diameter implies that the neighborhood graph is strongly connected. From the results it follows that to evaluate the neighborhood of a solution we only have to consider $\mathcal{O}(n^2)$ solutions and that we can reach any solution in $\mathcal{O}(n)$ steps. Compared to the $(n-1)!$ size of the complete solution space, these numbers are much more manageable. Note that a small diameter only indicates the *existence* of a small sequence of moves that results in a global optimum. Finding such a small sequence might still be difficult because of an unfavorable cost landscape. As such, a small diameter can be interpreted as a desirable property, which, however, is not sufficient for finding a (near) optimal solution fast.

Theorem 4.2. *The diameter of the node-insertion neighborhood graph for* TSP *is* $n-2$.

Proof. We first prove that the diameter d satisfies $d \leq n-2$ by showing for a pair τ, τ' of arbitrary tours that tour τ' can be obtained from τ by performing at most $n-2$ node-insertion moves. As tours are equivalent if they only differ in the city in which they start, we may assume that $\tau'(1) = \tau(1)$.

We construct τ' from τ by successively removing cities $\tau'(3), \tau'(4), \ldots, \tau'(n)$ from τ and inserting them in that order directly before $\tau'(1)$. If a city $\tau(i)$ is assigned to the same position as the position from which it was removed, then the node insertion on $\tau'(i)$ is skipped. As the total number of node insertions is at most $n-2$, we obtain $d \leq n-2$.

To prove $d \geq n-2$, assume that this is not true. So, let $d \leq n-3$. We show that this leads to a contradiction. Consider the tour $\tau = (1, 2, \ldots, n)$ and the tour $\tau' = (n, n-1, \ldots, 1)$ obtained from τ by reversing the order in which the cities are visited. By the assumption that $d \leq n-3$, tour τ' can be constructed from τ by performing at most $d \leq n-3$ node-insertion moves. This implies that τ' can be constructed by changing the position of at most $n-3$ cities and by keeping the mutual order of at least three cities intact. However, for any three cities i, j, k with $i < j < k$ we have that tour τ visits them in the order i, j, k, while τ' visits them in the order k, j, i. This gives rise to a contradiction. $\qquad\square$

Theorem 4.4 states that the 2-change neighborhood function also implies a neighborhood graph with an $\mathcal{O}(n)$ diameter. Recall that in Section 2.1 we have already indicated the following result.

Theorem 4.3. *For* TSP, *each node in the 2-change neighborhood graph has degree* $n(n-2)+2$. $\qquad\square$

Hence, the degree and diameter of the neighborhood graph of 2-change and node insertion is of comparable quality.

Theorem 4.4. *The diameter d of the 2-change neighborhood graph for* TSP *satisfies* $\frac{n}{2} \leq d \leq n-2$ *for any* $n \geq 5$.

Proof. Obviously, two tours are equivalent if they are equivalent in their first $n-1$ positions. Hence, to prove the upper bound on d, it suffices to prove by induction

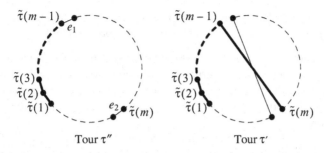

Tour τ'' Tour τ'

Figure 4.1. Definitions of e_1, e_2, and τ'. A bold edge indicates that the edge is also present in tour $\tilde{\tau}$, and a normal edge indicates that this is not necessarily the case.

on m that, for any two tours τ and $\tilde{\tau}$, we can transform τ into a tour τ' that satisfies $\tau'(i) = \tilde{\tau}(i)$ for $1 \leq i \leq m$ by performing a 2-change at most $m-1$ times. As motivated in the proof of Theorem 4.2, we may assume that $\tau(1) = \tilde{\tau}(1)$. Furthermore, without loss of generality we can consider SYMMETRIC TSP because a neighborhood contains by definition a tour in both possible directions. As a consequence, we may assume that $\tilde{\tau}(2)$ is one of the first $\lceil \frac{n}{2} \rceil$ cities in τ.

The induction hypothesis trivially holds for $m = 1$. Next, assume that $m > 1$. By the induction hypothesis, we can transform tour τ into a tour τ'' with $\tau''(i) = \tilde{\tau}(i)$ for $1 \leq i < m$ by performing a 2-change at most $m-2$ times. Hence, it suffices to show that if $\tau''(m) \neq \tilde{\tau}(m)$, then we can transform tour τ'' into a tour τ' satisfying $\tau'(i) = \tilde{\tau}(i)$ for $1 \leq i \leq m$ by performing a single 2-change. Let j be such that $\tau''(j) = \tilde{\tau}(m)$. Note that $m < j \leq n$. We construct τ' from τ'' by performing a 2-change on the edges $e_1 = \{\tau''(m-1), \tau''(m)\}$ and $e_2 = \{\tau''(j), \tau''(j+1)\}$; see Figure 4.1. These two edges are not successive as can be verified as follows. If e_1 were to directly precede e_2, then we would have $j = m$, which contradicts our assumption $\tau''(m) \neq \tilde{\tau}(m)$. Suppose, on the other hand, that e_1 directly succeeds e_2. In that case e_1 would be the first edge in tour τ'' and e_2 the last one. This would mean that $m = 2$ and that $\tilde{\tau}(2)$ is the last city in tour τ''. However, as $\tau'' = \tau$ for $m = 2$, this contradicts our assumption that $\tilde{\tau}(2)$ is one of the first $\lceil \frac{n}{2} \rceil$ cities in tour τ.

Because e_1 and e_2 are not successive, the 2-change based on these edges is feasible. Moreover, it does not change the subtour $\tilde{\tau}(1), \tilde{\tau}(2), \ldots, \tilde{\tau}(m-1)$ and it adds the edge $\{\tilde{\tau}(m-1), \tilde{\tau}(m)\}$ to the tour. Hence, tour τ' satisfies $\tau'(i) = \tilde{\tau}(i)$ for $1 \leq i \leq m$, which proves our induction hypothesis. This concludes the proof that $n-2$ is an upper bound on diameter d.

We now show that $\frac{n}{2}$ gives a lower bound on d. Let $\tau = (1, 2, \ldots, n)$. If n is odd, we define tour τ' as $(1, 3, 5, \ldots, n, 2, 4, 6, \ldots, n-1)$, i.e., it first visits all odd cities and then all even cities. If n is even, then we define τ' similarly, except that city n is visited between city 2 and 4. Hence, in that case we have that tour τ' is given by $\tau' = (1, 3, 5, \ldots, n-1, 2, n, 4, 6, \ldots, n-2)$. Clearly, tours τ and τ' do not have edges

in common. As a 2-change replaces at most two edges, this implies that at least $\frac{n}{2}$ 2-changes are necessary to construct τ' from τ. This proves $d \geq \frac{n}{2}$. \square

While in the above the focus was on studying the structure of individual neighborhood graphs, we now relate the structure of different neighborhood graphs. More specifically, we study the question of whether solutions that are neighbors of each other in one neighborhood function are also near to each other in another neighborhood function.

We start by relating 3-change and node insertion. An example of a node-insertion move is given in Figure 2.7. Obviously, the node insertion depicted in this figure is also a 3-change. The following theorem states that this is generally true. In the proof we write $suc_\tau(i)$ for the city that directly succeeds city i in tour τ. Analogously, $pred_\tau(i)$ defines the city that directly precedes city i in tour τ.

Theorem 4.5. *Each node insertion can be written as a 3-change. This means that the 3-change neighborhood function dominates*[1] *the node-insertion neighborhood function.*

Proof. Let τ' be obtained from τ by reinserting city i between city j and $suc_\tau(j)$. Then τ' is also obtained from τ by performing the 3-change that first replaces the edges $e_1 = \{pred_\tau(i), i\}$, $e_2 = \{i, suc_\tau(i)\}$, and $e_3 = \{j, suc_\tau(j)\}$ in the symmetric variant of τ by the edges $\{pred_\tau(i), suc_\tau(i)\}$, $\{j, i\}$, and $\{i, suc_\tau(j)\}$ and that next chooses the appropriate direction for the constructed symmetric tour. We note that if $pred_\tau(i) = suc_\tau(j)$, then the 3-change is actually a 2-change because the edge e_1 that is removed is the same as the edge $\{i, suc_\tau(j)\}$ that is added. \square

Note that the theorem implies that each local optimum of 3-change is also a local optimum of node insertion. Intuitively, the theorem states that when using 3-change we have a 'wider' view than with node insertion. This supports the better average-case performance that is obtained with the former neighborhood function. We note, however, that in individual cases node insertion might guide a given starting tour to a better local optimum.

The following theorem states that if tour τ' is a 3-change neighbor of tour τ, then τ' can be reached from τ in at most three steps in the 2-change neighborhood graph.

Theorem 4.6. *Let τ be an arbitrary tour and let τ' be a 3-change neighbor of τ. Then τ' can be constructed from τ by performing at most three 2-changes.*

Proof. Without loss of generality, we can restrict ourselves to SYMMETRIC TSP. By definition, τ' is obtained by either a single 2-change or by removing exactly three edges e_1, e_2, and e_3 from τ and replacing them by three new edges. We only have to consider the latter case. It can be verified that if all three edges selected from τ are successive, then we cannot construct a new tour by replacing them by three new edges. Hence, at least one selected edge from τ is not adjacent to the other two selected edges. First of all, assume that none of the three selected edges are adjacent. Then the only possibilities for tour τ' are alternatives 1 to 4 shown in Figure 4.2(a).

[1]See Definition 1.8.

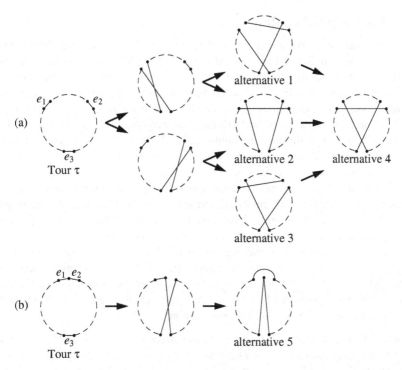

Figure 4.2. Construction of tour τ' from τ in at most three 2-changes in the case that (a) none of the three selected edges from τ are adjacent and (b) two of these edges are adjacent.

The figure also shows how the four alternatives can be constructed by a sequence of at most three 2-changes.

Next, assume that exactly two of the three selected edges from τ are adjacent. Figure 4.2(b) depicts the only possibility for tour τ' and it shows how τ' can be constructed from τ by performing two 2-changes. □

From Theorems 4.5 and 4.6 it follows that a node-insertion move can be simulated by at most three 2-changes. We can strengthen this claim to the following result.

Theorem 4.7. *Let τ be an arbitrary tour and let τ' be a node-insertion neighbor of τ. Then τ' can be constructed from τ by performing at most two 2-changes.*

Proof. In the proof of Theorem 4.5 we indicated that tour τ' can be constructed from τ by a 3-change on three edges of which two edges are adjacent. Furthermore, we showed in the proof of Theorem 4.6 that a 3-change on three edges of which two edges are adjacent can be simulated by two 2-changes. □

4.2 Job Shop Scheduling

In this section we revisit the indirect representation of JOB SHOP SCHEDULING
discussed in Section 3.3 and the critical-path swap neighborhood function we pro-
posed for it. It can easily be verified that, except for this neighborhood function, all
neighborhood functions discussed so far have the property that the corresponding
neighborhood graphs are strongly connected. This property means that if a local
search algorithm makes a long walk through the neighborhood graph and does not
stop at the first local optimum it encounters, then the algorithm can reach a globally
optimal solution as far as the structure of the neighborhood graph is concerned.

This property would also be true for JOB SHOP SCHEDULING if we would not
define the solution space as the set of machine orderings for which a feasible sched-
ule exists, but as the set of all machine orderings, and if we were to assume a neigh-
borhood function in which any two successive operations on the same machine may
be swapped instead of only two successive operations o and o' for which the arc
(o, o') is on a critical path in the digraph of the solution considered. Figure 3.8
shows that, as a result of the proposed restriction on both the solution space and
the neighborhood of a solution, the neighborhood graph is no longer strongly con-
nected. However, Theorem 4.8 states that it is still weakly optimally connected. To
prove this theorem, we need the following lemma.

Lemma 4.1. *Let Π be an arbitrary suboptimal machine ordering, and let Π^* be an
optimal machine ordering. Then D_Π contains a critical arc $(o, o') \in E_\Pi$ that does
not occur in D_{Π^*}, i.e., D_{Π^*} contains (o', o).*
Proof. Assume that the lemma does not hold. Hence, assume that each critical arc
in D_Π also occurs in D_{Π^*}. In that case the length of a longest path in Π^* is at least
the length of a longest path in Π, which is equivalent to saying that the cost of Π^* is
at least the cost of Π. This contradicts our assumption that Π is suboptimal and that
Π^* is optimal. □

Theorem 4.8. *The critical-path swap neighborhood graph for* JOB SHOP
SCHEDULING *is weakly optimally connected.*
Proof. We prove the theorem by contradiction. Let Π^* be an optimal machine
ordering and V be the non-empty set containing all machine orderings from which
we cannot reach an optimal machine ordering. Furthermore, let Π be a machine
ordering from V for which $\Delta(\Pi, \Pi^*)$ is minimal, where $\Delta(\Pi, \Pi^*)$ is the number
of arcs occurring in D_Π, but not in D_{Π^*}. Note that $\Delta(\Pi, \Pi^*) = \Delta(\Pi^*, \Pi)$. Ob-
viously, $\Delta(\Pi, \Pi^*) > 0$ holds, which by Theorem 3.3 and Lemma 4.1 implies that
we can construct a machine ordering Π' that is a neighbor of Π and for which
$\Delta(\Pi', \Pi^*) = \Delta(\Pi, \Pi^*) - 1$. As Π' must also be in V, this contradicts the property
that $\Delta(\Pi, \Pi^*)$ is minimal. □

A block of a machine ordering Π is defined as a maximal sequence of operations that
are processed consecutively and with no idle time on the same machine and that are
contained as a sequence in a longest path of D_Π. As mentioned in Section 3.3, the
critical-path swap neighborhood function is based on the observation that swapping

Figure 4.3. (a) Optimal schedule. (b) Schedule from which restricted neighborhood function cannot reach an optimal schedule.

the order of two successive operations o and o' on a machine can only result in a better solution if they are contained in a block. In addition, we have that such a swap can only be successful if o and o' are not both internal operations of a block. This can be verified as follows. Let o_{pred} be the predecessor of job o in a longest path p and let o'_{suc} be the successor of job o' in this path. As o and o' are internal operations, operations o_{pred}, o, o', and o'_{suc} are all processed on the same machine. This yields $(o_{pred}, o') \in E_{\Pi}$ and $(o, o'_{suc}) \in E_{\Pi}$. Hence, after swapping the order of o and o', i.e., after reversing $(o, o') \in E_{\Pi}$, the digraph contains path p with only the order of o and o' exchanged. This yields that the length of a longest path from s to t does not decrease. This observation suggests that we further restrict our neighborhood function, such that it only allows the reversal of critical arcs (o, o'), in which o and o' are not both internal operations of the same block. The following example shows that this restriction has the drawback that the neighborhood graph need no longer be weakly optimally connected.

Example 4.1. Assume we have four jobs, each consisting of two operations. The first operation of each job has a processing requirement of seven time units and it has to be processed on machine 1. The second operation of each job has to be processed on machine 2 and it has a processing requirement of two time units for jobs 1 and 2 and of one time unit for jobs 3 and 4. Figure 4.3(a) depicts an optimal schedule for this problem instance. The schedule has a makespan of 29.

As the total processing time on machine 2 is smaller than the processing requirement of a single operation on machine 1, it holds for any machine ordering $\Pi = (\pi_1, \pi_2)$ that the longest path in digraph D_{Π} starts with the block B_1 given by $B_1 = \pi_1(1), \pi_1(2), \pi_1(3), \pi_1(4)$ and ends with a block B_2 containing at least the second operation o_{2j} of the job j containing operation $\pi_1(4)$. Based on this, we can make the following two observations. (*i*) The makespan of an optimal schedule for

Π is at least $4 \cdot 7 = 28$ plus the processing requirement p_{2j} of operation o_{2j}; see Figures 4.3(a) and 4.3(b). (*ii*) Regarding machine 1, the new neighborhood function only allows us to swap either the first two or the last two operations. From the first observation it follows that, in an optimal schedule, the last job processed by machine 1 is either job 3 or job 4. From the second observation it follows that if we start with the machine ordering corresponding to the schedule given in Figure 4.3(b), then we can never reach a machine ordering in which machine 1 ends with processing job 3 or job 4. When we combine these two conclusions we obtain that we cannot reach an optimal solution from the machine ordering underlying the schedule of Figure 4.3(b). □

4.3 Bibliographical Notes

Theorem 4.4 states that the diameter of the 2-change neighborhood graph is bounded between $\frac{n}{2}$ and $n-2$. Stadler & Schnabl [1992] prove the slightly weaker result that this diameter is bounded between $\frac{n}{2}$ and $n-1$.

Section 4.2 is based on the work of Van Laarhoven [1988] and Van Laarhoven, Aarts & Lenstra [1992]. Matsuo, Suh & Sullivan [1988] observe that reversing a critical arc (i, j) cannot improve a solution if both i and j are internal operations of a block. By a different and slightly more complicated example than Example 4.1 Dell'Amico & Trubian [1993] show that using this observation yields a neighborhood function that is no longer weakly optimally connected.

4.4 Exercises

1. Show that the node-insertion neighborhood graph has a smaller diameter for SYMMETRIC TSP than for TSP, which by Theorem 4.2 is $n-2$.

2. Consider TSP. We define the inversion-and-insertion neighborhood function, such that a tour τ' is a neighbor of tour τ if and only if it can be obtained by removing a segment of any number of consecutive cities from τ and reinserting the segment as is or inverted between any two remaining cities in τ.

 a) Prove that this neighborhood function is equivalent to the 3-change neighborhood function.

 b) What does this result say about the relation between the 3-change neighborhood function and the Or neighborhood function?

3. Let $G = (V, E)$ be a graph with maximum degree Δ and with diameter D.

 a) Prove that

 $$|V| \le 1 + \sum_{i=0}^{D-1} \Delta(\Delta - 1)^i. \tag{4.1}$$

 b) Show that (4.1) can be rewritten to

 $$|V| \le \frac{\Delta(\Delta - 1)^D - 2}{\Delta - 2}.$$

 This bound is known as the Moore bound.

c) Use the Moore bound to derive a lower bound on the diameter of the 2-change neighborhood graph for TSP with $n = 8, 10$, and 12 cities.

4. We define graph $Q_n = (V, E)$, such that the node set V is given by the set of binary sequences of length n and $\{i, j\} \in E$ for $i, j \in V$ if and only if the Hamming distance between i and j is 1. Graph Q_n is called a hypercube of dimension n.

 a) Give the degree and diameter of hypercube Q_n.

 b) Prove an upper bound on the number of local optima in a neighborhood graph that is given by Q_n.

5

Performance Guarantees

The two main aspects that determine the performance of an algorithm are the quality of the solutions it finds and its running time. In this chapter we study the former for iterative improvement, which corresponds to determining the quality of local optima. The running time of iterative improvement and, more generally, the complexity of finding a local optimum is the subject of Chapter 6.

To measure the quality of the solutions returned by a given algorithm, we can perform the following types of analyses.

- *Empirical analysis.* The performance of the algorithm is evaluated over a possibly large set of problem instances.

- *Probabilistic analysis.* This analysis aims to determine the average-case performance for a given probability distribution of the instances.

- *Worst-case analysis.* Here, we are interested in the worst case, i.e., guaranteed, performance of the algorithm over all possible problem instances.

In most cases a local search algorithm is evaluated by means of an empirical analysis. We focus on the worst-case performance of iterative improvement.

In Section 5.1 we give examples in which iterative improvement is guaranteed to end up in an optimal solution, which means that the underlying neighborhood function is exact. Next, we analyze the performance ratios of local optima for the k-change neighborhood function and for some machine-scheduling neighborhood functions in Section 5.2. We note that a neighborhood function has a performance bound U if all its local optima have a cost at most U times the optimal cost. If

bound U is tight, then U is called a performance ratio. The terms performance ratio
and performance bound are also used to indicate the quality of an algorithm; see
Appendix B. In Sections 5.3 and 5.4 we finally elaborate on the relation between
the formal hardness of finding an R-approximate solution and the existence of an
efficient neighborhood function with a performance bound of R.

5.1 Exact Neighborhood Functions

A neighborhood function is said to be exact if each locally optimal solution is also
globally optimal. In this section we present some examples of exact neighbor-
hood functions. First of all, consider SORTING. Sorting a multiset of n numbers
a_1, a_2, \ldots, a_n corresponds to finding a permutation π of $\{1, 2, \ldots, n\}$ that minimizes
the cost function $f(\pi) = \sum_{i=1}^{n} i \cdot a_{\pi(i)}$ if the numbers have to be put in non-increasing
order and that maximizes this cost function if they have to be put in non-decreasing
order. A well-known algorithm for SORTING is bubble sort. This algorithm starts
with an arbitrary permutation π and then swaps repeatedly two adjacent numbers
that are not in the right order. Bubble sort finds an optimal solution to SORTING be-
cause a sequence of numbers is sorted if and only if any two adjacent numbers in the
sequence are sorted correctly. Obviously, swapping two adjacent numbers improves
the cost of a permutation if and only if they are not in the right order. This means
that bubble sort corresponds to iterative improvement with the neighborhood func-
tion that constructs neighbors by swapping two adjacent numbers. Because bubble
sort solves SORTING, this neighborhood function is exact.

The swap neighborhood function is also exact for the machine scheduling prob-
lem $1||\sum_j w_j C_j$. For this problem Theorem 3.1 states that an optimal schedule is
obtained by putting the jobs in SWPT order. In the proof of Theorem 3.1, we ar-
gued that we can improve upon a suboptimal schedule by swapping two adjacent
jobs. Hence, the only schedules for which no improving swap exists are the opti-
mal schedules. This implies that the swap neighborhood function is exact. Note
that although this machine scheduling problem is equivalent to sorting a multiset
of numbers in non-decreasing order, the exactness of the swap neighborhood func-
tion does not follow directly from the exactness of this neighborhood function for
SORTING because we consider different cost functions for the problems.

In Section 2.3 we presented a polynomial-time algorithm for MINIMUM SPAN-
NING TREE, in which we are asked to find a spanning tree with minimum weight
for an edge-weighted graph. This problem can also be solved by applying itera-
tive improvement with the exact neighborhood function proposed in the following
theorem.

Theorem 5.1. *Consider for* MINIMUM SPANNING TREE *the neighborhood function
that generates a neighbor of a spanning tree T in the following way. First of all, we
add an arbitrary edge to T. This produces a single cycle. From this cycle we delete
an arbitrary edge, which gives a new spanning tree. This neighborhood function is
exact.*

Proof. Suppose we have a spanning tree T that is a local optimum, but not a global

Figure 5.1. Convex hull of the French cities Paris, Strasbourg, Lyon, Nice, Marseille, Toulouse, Bordeaux, Nantes, Le Havre, and Lille. Paris and Lyon are interior cities.

optimum. To prove the theorem, we show that this yields a contradiction. Let T^* be an optimal spanning tree that has a maximum number of edges in common with T. Furthermore, let e be an edge with minimum weight from the set of edges that are in T^* but not in T. As any spanning tree has $n-1$ edges, such an edge exists. The removal of e splits T^* into two subtrees T_1^* and T_2^*. We now construct a neighbor T' of T as follows. First, we add edge e to T. This results in a cycle c. Besides e, this cycle must contain at least one other edge e' connecting a node from T_1^* with a node from T_2^*. We now construct T' by removing e' from c.

Because T is a local optimum, we have that the cost of T' is at least the cost of T. Hence, $w(e) \geq w(e')$, where $w(e)$ and $w(e')$ give the weights of edges e and e'. However, this implies that connecting T_1^* and T_2^* by edge e' results in a spanning tree that is at least as good as T^* and that has at least one edge more in common with T. This yields a contradiction. Hence, the proposed neighborhood function is exact. □

5.1.1 *Euclidean TSP with Few Interior Cities

Consider an arbitrary problem instance of EUCLIDEAN TSP. The convex hull of its n cities is defined as the smallest convex polytope that contains all n cities. A city is called interior if it is not on the boundary of the convex hull. Figure 5.1 depicts a problem instance of EUCLIDEAN TSP with ten cities, of which two cities are interior. We prove that 2-change is exact for problem instances of EUCLIDEAN TSP with no interior cities. For problem instances with $n_{\text{in}} \geq 1$ interior cities, we show that k-change is exact if and only if $k \geq 2n_{\text{in}} + 1$. For the example of Figure 5.1, the result implies that 5-change is guaranteed to give an optimal tour.

In the proof of these results, we use that locally optimal tours have no touching, intersecting, or overlapping edges. We first prove this property.

Definition 5.1. Consider an arbitrary tour for a problem instance of EUCLIDEAN TSP, and let e_1 and e_2 be two edges in this tour. If e_1 and e_2 have infinitely many points in common, then they are called *overlapping*. If e_1 and e_2 only have one

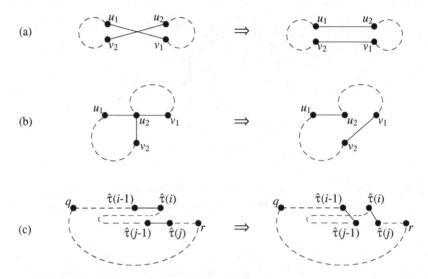

Figure 5.2. Cases distinguished in the proof of Lemma 5.1. The cities between q and r are lying on the edge $\{q,r\}$.

point p in common, then they are called *intersecting* in the case where p is not an end point of at least one of the edges and they are said to *touch* in the case where p is an end point of exactly one of the two edges. ☐

We note that it follows from the definition above that if two edges are successive, they can possibly overlap, but they cannot intersect or touch.

Lemma 5.1. *Consider a problem instance of* EUCLIDEAN TSP, *in which not all cities lie on a straight line. Furthermore, let $\hat{\tau}$ be a locally optimal tour with respect to the k-change neighborhood function for any $k \geq 2$. Then two arbitrary edges in $\hat{\tau}$ do not touch, intersect, or overlap.*

Proof. As a 2-change is also a k-change for any $k \geq 3$, it suffices to prove that the lemma holds for $k = 2$. Let $\hat{\tau}$ be a locally optimal tour with respect to the 2-change neighborhood function, and let $\hat{\tau}$ contain two edges that intersect, touch, or overlap. To prove the lemma, we show that this yields a contradiction.

First, suppose that $\hat{\tau}$ contains two intersecting edges. We then have the situation sketched in Figure 5.2(a) for some pair of non-successive edges $e_1 = \{u_1, v_1\}$ and $e_2 = \{u_2, v_2\}$. Let p be the intersection point of e_1 and e_2. Obviously, we have $d_{u_1,p} + d_{p,u_2} > d_{u_1,u_2}$ and $d_{v_1,p} + d_{p,v_2} > d_{v_1,v_2}$ (note that this does not follow directly from the triangle inequality, which states that $d_{u_1,p} + d_{p,u_2} \geq d_{u_1,u_2}$ and $d_{v_1,p} + d_{p,v_2} \geq d_{v_1,v_2}$). As a result, replacing e_1 and e_2 by $\{u_1, u_2\}$ and $\{v_1, v_2\}$ improves the length of tour $\hat{\tau}$, which contradicts our assumption that $\hat{\tau}$ is a local optimum.

Next, assume that $\hat{\tau}$ contains two edges $e_1 = \{u_1, v_1\}$ and $e_2 = \{u_2, v_2\}$ that

touch; see Figure 5.2(b). Also in this case, replacing e_1 and e_2 by $\{u_1, u_2\}$ and $\{v_1, v_2\}$ yields a tour that is shorter than $\hat{\tau}$. Hence, we again arrive at a contradiction.

Finally, suppose that the locally optimal tour $\hat{\tau}$ contains two overlapping edges e_1 and e_2 lying on a line l. We define cities q_i and r_i with $i = 1, 2$, such that the tour enters l at q_i, remains on l until it visits e_i, and leaves l for the first time at r_i. The part of the tour that starts at q_i and ends at r_i is called a section. All cities visited in this section must lie between q_i and r_i. Otherwise, the edge entering q_i or leaving r_i touches the edge in the section passing q_i or r_i, which yields a contradiction, as we showed above.

If e_1 and e_2 do not belong to the same section, but to two different sections s_1 and s_2, then at least one of the two edges adjacent to s_1 touches an edge of s_2 or vice versa. This means that we may assume that e_1 and e_2 belong to the same section. As a result, we can assume that, between visiting $q = q_1 = q_2$ and $r = r_1 = r_2$, tour $\hat{\tau}$ moves back and forth on line l at least once. From this assumption, we will derive a contradiction.

Let V be the set of cities contained in section $s_1 = s_2$, and let the cities in V be numbered from 1 to $|V|$ in order of increasing distance from q. Without loss of generality we assume $\hat{\tau}(1) = 1$, i.e., q is the first city in tour $\hat{\tau}$. By assumption, $\hat{\tau}$ moves back and forth on line l when visiting the cities from V. This means that an $i < |V|$ exists with $\hat{\tau}(i) > \hat{\tau}(i+1)$. Let i be the smallest integer with this property. Furthermore, let j be the smallest integer larger than i with $\hat{\tau}(j) > \hat{\tau}(i)$; see Figure 5.2(c). As city $|V|$, i.e., city r, is visited last, j exists. Note that $i \geq 2$ and $j \geq i+2$. Performing a 2-change on the edges $\{\hat{\tau}(i-1), \hat{\tau}(i)\}$ and $\{\hat{\tau}(j-1), \hat{\tau}(j)\}$ yields a neighboring tour τ of $\hat{\tau}$. We now show that τ is shorter than $\hat{\tau}$, which contradicts our assumption that $\hat{\tau}$ is locally optimal.

Let c be the smaller of the two cities $\tau(i-1)$ and $\tau(j-1)$. By the minimality of i, we have $\hat{\tau}(i-1) < \hat{\tau}(i)$. Furthermore, by the minimality of j and because $j \geq i+2$, we have $\hat{\tau}(j-1) < \hat{\tau}(i)$, which yields $\hat{\tau}(j-1) < \hat{\tau}(i) < \hat{\tau}(j)$. As a result, we obtain $d_{c,\hat{\tau}(j)} < d_{\hat{\tau}(i-1),\hat{\tau}(i)} + d_{\hat{\tau}(j-1),\hat{\tau}(j)}$ and $d_{c,\hat{\tau}(j)} > d_{\hat{\tau}(i-1),\hat{\tau}(j-1)} + d_{\hat{\tau}(i),\hat{\tau}(j)}$. This implies that the length of tour τ minus the length of tour $\hat{\tau}$, which is given by $d_{\hat{\tau}(i-1),\hat{\tau}(j-1)} + d_{\hat{\tau}(i),\hat{\tau}(j)} - d_{\hat{\tau}(i-1),\hat{\tau}(i)} - d_{\hat{\tau}(j-1),\hat{\tau}(j)}$, is negative. Hence, τ is indeed shorter than $\hat{\tau}$. This proves the lemma. □

We can now prove the result claimed for the special case where we do not have any interior cities.

Theorem 5.2. *The 2-change neighborhood function is exact for problem instances of* EUCLIDEAN TSP *with no interior cities. It is not exact for the set of problem instances of* EUCLIDEAN TSP *with one interior city.*

Proof. We first prove that 2-change is exact for an arbitrary problem instance of EUCLIDEAN TSP that does not contain interior cities. Consider the case in which not all cities lie on a straight line. Then the tours that visit all cities in either clockwise or counterclockwise order are the only tours in which any two edges do not touch, intersect, or overlap. Hence, as the starting position and the direction of a tour are irrelevant, Lemma 5.1 implies that we only have one local optimum. Obviously, this

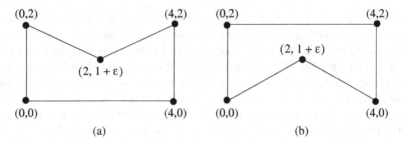

Figure 5.3. Example showing that 2-change does not have to be exact if we are considering problem instances of EUCLIDEAN TSP with one interior city. Tour (a) is globally optimal, while (b) is only locally optimal.

tour also has to be a global optimum. This yields that the neighborhood function is exact.

Next, assume that all n cities are on a straight line l. In the proof of Lemma 5.1 we showed that, between visiting any two cities, a locally optimal tour cannot move back and forth on the line segment between them. Hence, a locally optimal tour first moves in a straight line from the innermost city to the outermost city and then moves in a straight line from the outermost city to the innermost city. This clearly results in an optimal tour length. Consequently, we find that the 2-change neighborhood function is also exact if all cities lie on a straight line.

We now prove the second part of the theorem by giving a problem instance with one interior city for which 2-change is not exact. Figure 5.3 shows a problem instance with five cities. The figure gives the positions of the cities in the two-dimensional coordinate system. The four cities $(0,0), (4,0), (4,2)$, and $(0,2)$ determine the convex hull and all lie on integer grid points. The only interior city $(2, 1 + \varepsilon)$ lies slightly above the middle of the convex hull. Consider the two tours given in Figures 5.3(a) and 5.3(b). It can be verified that tour (a) is shorter than tour (b). Furthermore, it can be verified that all other tours are worse than these two tours. Hence, as we require three edges to construct tour (b) from (a), tour (b) is a local optimum, but not a global optimum with respect to 2-change. This shows that 2-change is not exact for the set of problem instances of EUCLIDEAN TSP with one interior city. □

We now focus on the case in which we have at least one interior city, i.e., $n_{in} \geq 1$. To prove that $(2n_{in} + 1)$-change is exact, we use an auxiliary result. This result, which is proved below, is formulated as follows. Let P_1 and P_2 be two partitions of a set V of items. If P_1 and P_2 are 'non-separable', then the total number of subsets in P_1 and P_2 does not exceed $|V| + 1$. Partitions P_1 and P_2 are said to be non-separable if no subset $V' \subset V$ exists that is partitioned by a selection of the subsets of P_1 as well as by a selection of the subsets of P_2. In other words, P_1 and P_2 are non-separable if no subset $V' \subset V$ exists such that each subset in P_1 as well as in P_2 is completely either

in V' or in $V \setminus V'$. To illustrate this, let $V = \{1, 2, 3\}$ and let partition P_1 contain the two subsets $\{1, 2\}$ and $\{3\}$ and P_2 the three subsets $\{1\}$, $\{2\}$, and $\{3\}$. Then P_1 and P_2 are separable as, for instance, $V' = \{1, 2\}$ is partitioned by P_1' containing the first subset $\{1, 2\}$ of P_1 and by P_2' containing the first two subsets $\{1\}$ and $\{2\}$ of P_2. However, the partitions P_1 and P_3, where P_3 is given by $\{1\}$ and $\{2, 3\}$, are non-separable, as can be verified.

Definition 5.2. Let P_1 and P_2 be two partitions of a set V of items. Furthermore, let graph $G = (V, E_1 \cup E_2)$ be such that $\{u, v\} \in E_i$ with $i \in \{1, 2\}$ if and only if u and v are assigned to the same subset in partition P_i. Then P_1 and P_2 are said to be *non-separable* if G is connected. □

Lemma 5.2. *Let P_1 and P_2 be two partitions of $V = \{1, 2, \dots, n_V\}$. Furthermore, let m_i be the number of subsets in P_i for $i \in \{1, 2\}$. If P_1 and P_2 are non-separable, then we have $m_1 + m_2 \leq n_V + 1$.*

Proof. Assume that P_1 and P_2 are non-separable, and let $G = (V, E_1 \cup E_2)$ be the connected graph induced by the two partitions. We prove that $m_1 + m_2 \leq n_V + 1$ holds.

When we leave the edges from E_2 out of consideration, graph G consists of m_1 components. Let G' be defined as the graph whose nodes correspond to these m_1 components and which contains an edge between components c and c' with $c \neq c'$ if G contains an edge $\{j, j'\} \in E_2$ between items $j \in c$ and $j' \in c'$. Since G is connected, G' is also connected. Let T be an arbitrary spanning tree of G'. We choose one of its nodes as its root, and we direct all its edges from the root to the leaf nodes. Obviously, T contains $m_1 - 1$ arcs.

We now perform the following labeling action on each arc $a = (c, c')$ in T. We select one edge $\{j, j'\} \in E_2$ with $j \in c$ and $j' \in c'$ corresponding to arc a. We then label item j', i.e., the item corresponding to the head of arc a.

For the derived labeling, the following two properties hold. In the first place, we have that zero items are labeled from the component represented by the root node and exactly one item is labeled from a component that is represented by some other node, where this single item is only labeled once. This is true because in directed tree T the root has indegree zero and each other node has indegree one. Secondly, we have that each subset in P_2 contains at least one unlabeled item. To show that this claim is true, suppose that this is not the case and let A be a subset in P_2 containing only labeled items. Furthermore, let W be the set of components in T that contain an item from A. Since all nodes in A are labeled, it follows that T contains for each $c \in W$ an arc that is directed from some node in W to node c. However, this implies that T contains a cycle, which contradicts the fact that it is a tree. This proves our second property.

From the second property, it follows that m_2 is at most the number of unlabeled items, which by the first property is given by $n_V - (m_1 - 1)$. Hence, we obtain $m_2 \leq n_V - m_1 + 1$, which shows that indeed $m_1 + m_2 \leq n_V + 1$. □

Having proved Lemma 5.2 we are now ready to prove the following result.

interior chain

Figure 5.4. Example of a tour with four interior chains.

Theorem 5.3. *The k-change neighborhood function is exact for the set of problem instances of* EUCLIDEAN TSP *with* $n_{in} \geq 1$ *interior cities if and only if* $k \geq 2n_{in} + 1$.
Proof. We first prove that the $(2n_{in} + 1)$-change neighborhood function is exact. Assume we have n_{in} interior cities with $n_{in} \geq 1$. Note that this implies that not all cities lie on a straight line. Furthermore, let $\hat{\tau}$ be a locally optimal tour with respect to $(2n_{in} + 1)$-change and let τ^* be a shorter globally optimal tour. We show that this leads to a contradiction.

We define an interior chain of a tour as a sequence of successive interior cities in the tour. An interior chain is maximal in the sense that both the predecessor of the first city and the successor of the last city are on the boundary of the convex hull; see Figure 5.4. From Lemma 5.1, it follows that both $\hat{\tau}$ and τ^* visit the cities on the boundary of the convex hull in either clockwise or counterclockwise order. Without loss of generality, we assume that the direction is the same for both $\hat{\tau}$ and τ^*. Hence, tours $\hat{\tau}$ and τ^* are determined completely by their interior chains and, for each interior chain, the city on the boundary that precedes the chain.

The interior chains of a tour define a partition of the n_{in} interior cities, such that two cities belong to the same subset if and only if they are in the same interior chain. Let \hat{P} and P^* be the partitions implied by $\hat{\tau}$ and τ^*, respectively. Furthermore, let graph $G = (V, E_1 \cup E_2)$ be defined as in Definition 5.2, where V is the set of interior cities, E_1 refers to \hat{P}, and E_2 refers to P^*.

Consider an arbitrary component C of G. We define V_C as the set of nodes in C. By definition, the edges in C from E_1 completely define for $\hat{\tau}$ which cities from V_C are contained in which interior chain. Similarly, the edges from E_2 completely define for τ^* which cities from V_C are contained in which interior chain. We show that by a single $(2n_{in} + 1)$-change, we can transform the interior chains covering V_C in $\hat{\tau}$ into the interior chains on V_C in τ^* without affecting the interior chains defined by any other component in G and the way these chains are connected to the tour. Obviously, because τ^* is shorter than $\hat{\tau}$, at least one component in G must exist for which this transformation yields an improvement. However, the existence of such a transformation contradicts our assumption that $\hat{\tau}$ is locally optimal. Hence, to prove the exactness of the $(2n_{in} + 1)$-change neighborhood function, it suffices to prove that we can transform the interior chains on V_C in locally optimal tour $\hat{\tau}$ into the interior chains on V_C in globally optimal tour τ^*.

Let $\hat{P}_C \subseteq \hat{P}$ and $P_C^* \subseteq P^*$ be the partitions defined by G for the cities in V_C. Hence, \hat{P}_C defines the \hat{N}_{ic} interior chains covering V_C in $\hat{\tau}$ and P_C^* defines the N_{ic}^* interior chains covering V_C in τ^*. Disconnecting all interior cities from V_C in $\hat{\tau}$ requires the removal of $\hat{N}_{ic} + |V_C|$ edges. We can now construct the interior chains covering V_C as they are given in τ^*. To insert these chains in tour $\hat{\tau}$ in the way this is done in τ^*, we have to remove at most N_{ic}^* additional edges from the boundary of the convex hull. Below we prove that it is possible to insert an interior chain at the right position. This means that if we have to insert an interior chain after city i, then no interior chain that contains interior cities outside V_C has already been inserted directly after city i. Hence, by replacing $\hat{N}_{ic} + |V_C| + N_{ic}^*$ edges, we can replace the interior chains covering V_C in $\hat{\tau}$ by the interior chains covering V_C in $\hat{\tau}$. Using Lemma 5.2, we obtain $\hat{N}_{ic} + N_{ic}^* \leq |V_C| + 1$. As a result, we have to replace at most $2|V_C| + 1 \leq 2n_{in} + 1$ edges, which can be done by the $(2n_{in} + 1)$-neighborhood function.

Hence, all that remains to be shown is that if in τ^* an interior chain c^* with nodes from V_C has been inserted directly after city i on the boundary of the convex hull and if $\hat{\tau}$ also has an interior chain \hat{c} that starts directly after city i, then interior chain \hat{c} only contains cities from V_C. Assume that this is not the case. By the definition of V_C, this means that none of the cities of \hat{c} belong to V_C. Let j be the city that succeeds i on the boundary of the convex hull in both $\hat{\tau}$ and τ^*. Note that j directly succeeds \hat{c} and c^*. We define $h(c^*)$ as the hull obtained by adding edge $\{i, j\}$ to c^* and $h(\hat{c})$ as the hull obtained by adding edge $\{i, j\}$ to \hat{c}. We have that at least one of the following cases holds: (i) at least one city of \hat{c} is in or on $h(c^*)$ and (ii) at least one city of c^* is in or on $h(\hat{c})$. However, both cases lead to a contradiction. By Lemma 5.1, the first case contradicts the global optimality of τ^*, while the second case contradicts the local optimality of $\hat{\tau}$. This completes the proof that the $(2n_{in} + 1)$-change neighborhood function is exact.

The claim that $2n_{in}$-change does not have to be exact for EUCLIDEAN TSP with at most n_{in} interior cities can be proved by example. Assume that n_{in} is even and consider the problem instance given in Figure 5.5. Unless indicated otherwise, all edges have length one and the upper left corner and lower left corner are both right angles. Furthermore, M is chosen relatively large and $\varepsilon > 0$ very small. The values of δ and M' can be derived from M and ε. It can be verified that the tour given in Figure 5.5(a) is a locally optimal tour with respect to $2n_{in}$-change. However, the globally optimal tour given in Figure 5.5(b) is ε shorter. Based on similar ideas, we can also construct a problem instance for which $2n_{in}$-change is not exact in the case where n_{in} is odd. □

5.2 Performance Ratios of Neighborhood Functions

For many problems an efficient exact neighborhood function is unattainable. In this section we analyze the quality of solutions returned by iterative improvement for some neighborhood functions that are not exact. More specifically, we study for these neighborhood functions the quality of local optima relative to that of global optima.

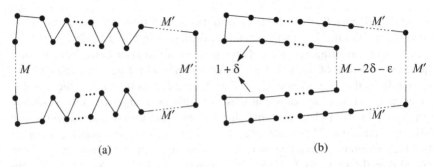

<center>(a) (b)</center>

Figure 5.5. Example showing that $2n_{\text{in}}$-change does not have to be exact if we are considering problem instances of EUCLIDEAN TSP with n_{in} interior city, where $n_{\text{in}} > 0$ is even. Tour (a) is locally optimal, and tour (b) is globally optimal.

Section 5.2.1 is devoted to analyzing the performance ratio of the k-change neighborhood function for SYMMETRIC TSP and METRIC TSP, while in Section 5.2.2 we analyze the performance ratio of the move and swap neighborhood functions for the machine scheduling problems $Pm||C_{\max}$ and $Qm||C_{\max}$.

5.2.1 Performance Ratio of the k-Change Neighborhood Function

Consider SYMMETRIC TSP. As the neighborhood size of k-change is polynomially bounded, a single iteration of k-Opt, which corresponds to a single step in the transition graph, can be implemented to run in polynomial time. Furthermore, Theorem 2.1 states that finding a tour with length at most $2^{p(n)}$ times the optimal length is NP-hard for any fixed polynomial p. Hence, if we were to prove that k-Opt is guaranteed to reach a locally optimal tour within a polynomial number of iterations, then we would obtain that, unless P = NP, k-Opt does not admit a performance ratio at most $2^{p(n)}$ for any fixed polynomial p. However, as we show in the next chapter, k-Opt may require an exponential number of iterations, which implies that the given reasoning is not applicable. The following theorem states that the lower bound of $2^{p(n)}$ on the performance ratio of k-Opt nevertheless holds. It even claims a stronger result: no function in n exists that gives an upper bound on the performance ratio of k-change. We note that this does not mean that it is impossible to bound the performance ratio. It can, for instance, be bounded trivially by the largest distance in the distance matrix d divided by the smallest distance in d.

Theorem 5.4. *For any fixed $k \geq 2$ and $n \geq 2k + 4$, the performance ratio of the k-change neighborhood function cannot be bounded by a constant for SYMMETRIC TSP on n cities. This means that no function in n exists that gives a performance bound for k-change.*

Proof. For any $k \geq 2$ and $\varepsilon > 0$, we define a problem instance of SYMMETRIC TSP containing $n = 2k + 4$ cities. For this problem instance we construct a locally optimal tour $\hat{\tau}$, which has a performance ratio of $1 + \frac{1}{n\varepsilon}$. It is easy to extend this problem instance and to modify the locally optimal tour, such that the problem instance con-

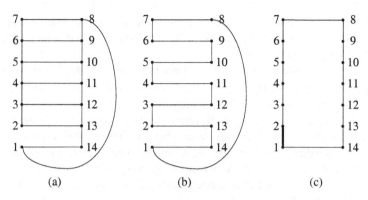

Figure 5.6. For $k = 5$, the figure shows (a) graph G, (b) optimal tour τ^* of length $n\varepsilon$, and (c) locally optimal tour $\hat{\tau}$ of length $1 + n\varepsilon$.

tains any $n > 2k + 4$ cities, the tour $\hat{\tau}$ remains locally optimal, and the performance ratio of $\hat{\tau}$ is still $1 + \frac{1}{n\varepsilon}$. Since for $\varepsilon \to 0$ this performance ratio approaches ∞, this proves the theorem.

We define the symmetric distance matrix d, such that all entries are given by $1 + \varepsilon$ except for the entries $d_{i,i+1}$ with $1 < i < n$, $d_{1,n}$, $d_{1,k+2+l}$ with $l = k \bmod 2$, and $d_{i,n-i+1}$ with $2 \leq i \leq k + 2$, which are all given by ε. In this definition we restrict ourselves without loss of generality to the entries d_{ij} with $i < j$.

Let $G = (V, E)$ be the graph obtained from the complete graph of n nodes by maintaining an edge $\{i, j\}$ if $d_{ij} = \varepsilon$ and by removing it if $d_{ij} = 1 + \varepsilon$; see Figure 5.6. Obviously, a tour has length $n\varepsilon$ if it defines a Hamiltonian cycle in graph G and it has length at least $1 + n\varepsilon$ otherwise. Moreover, it can be proved that the tour $\tau^* = (1, n, n - 1, 2, 3, n - 2, n - 3, \ldots)$ defines the only Hamiltonian cycle in G, where tours that only differ in their starting position or direction are considered to be equivalent. Hence, tour τ^* is the only optimal tour and the length of any other tour τ is at least $1 + n\varepsilon$. From this it follows that if a tour $\hat{\tau}$ has length $1 + n\varepsilon$ and if it has $k + 1$ edges that are absent in τ^*, then $\hat{\tau}$ is a locally optimal tour with performance ratio $\frac{1+n\varepsilon}{n\varepsilon} = 1 + \frac{1}{n\varepsilon}$. It can be verified that such a tour is given by $\hat{\tau} = (1, 2, \ldots, n)$. This proves the theorem. $\qquad\square$

In the remainder of this section, we focus on METRIC TSP. Although solving this problem is as difficult as solving SYMMETRIC TSP (they are both strongly NP-hard), it is easier to find an approximate solution for METRIC TSP than for SYMMETRIC TSP, as is stated by Theorem 2.1. Currently, Christofides' algorithm is the polynomial-time algorithm with the best known worst-case performance. This constructive algorithm works as follows. First of all, it computes a minimum spanning tree T. Next, it converts this tree into an Eulerian graph, which is a graph in which each node has an even degree. This is done by deriving a perfect matching on the nodes with an odd degree. These edges are added to T. For an Eulerian graph, we can easily determine an Eulerian tour, which is a tour that traverses each edge

exactly once. By introducing shortcuts, we can eliminate multiple visits to the same city in the tour. This results in a tour for which it can be proved that its length is at most $\frac{3}{2}$ times the optimal length.

For k-Opt it has been proved that, for infinitely many n, a lower bound on the performance ratio is given by $\frac{1}{4}\sqrt{n}$ for $k = 2$ and by $\frac{1}{4}n^{\frac{1}{2k}}$ for $k \geq 3$. Hence, from a worst-case perspective the local search algorithm k-Opt cannot compete against Christofides' constructive algorithm. Nevertheless, k-Opt may still be preferred over Christofides' algorithm as in practice we may be more interested in a good average-case performance than in a good worst-case performance. Hence, it is still interesting to analyze the worst-case performance of k-Opt or, equivalently, k-change, not only to obtain a worst-case performance guarantee, but also to enhance our understanding of the algorithm.

We will prove that k-Opt has a performance bound of $4\sqrt{n}$ for any $k \geq 2$. Note that this performance bound differs by a constant factor from the presented lower bound for $k = 2$ and by a factor that increases with n for any $k \geq 3$. In order to prove the result, we need the following auxiliary result that states that a locally optimal tour cannot contain too many long edges as expressed as a fraction of the optimal tour length.

Lemma 5.3. *For an arbitrary problem instance of* METRIC TSP, *let* l^* *be the optimal tour length and* \hat{t} *a locally optimal tour with respect to the 2-change neighborhood function. For any* i *with* $1 \leq i \leq n$ *we have* $|E_i| < i$, *where set* E_i *contains the edges* $\{j, j'\}$ *from* \hat{t} *with* $d_{jj'} > \frac{2l^*}{\sqrt{i}}$.

Proof. We prove the lemma by contradiction. So, suppose we have $|E_i| \geq i$ for some i. Although the direction of \hat{t} is irrelevant, we assume some orientation of the tour. As a result, the $r = |E_i|$ edges in E_i transform into r arcs (t_j, h_j) with $1 \leq j \leq r$. The cities t_j and h_j are called the tail and head of arc (t_j, h_j).

We first show that a tail t_u with $1 \leq u \leq r$ cannot be surrounded by too many other tails of arcs from E_i. Let V_u be the set of all tails of arcs from E_i that have a distance of at most $\frac{l^*}{\sqrt{i}}$ from t_u. Assume that $|V_u| \geq \sqrt{i}$. We prove that $|V_u| < \sqrt{i}$ by showing that this assumption yields a contradiction.

Let $t_j, t_{j'} \in V_u$. Because the triangle inequality holds and because, by definition, the distance from t_u to any of these two tails is at most $\frac{l^*}{\sqrt{i}}$, we obtain that the distance between t_j and $t_{j'}$ is at most $\frac{2l^*}{\sqrt{i}}$. This implies that the distance between the heads h_j and $h_{j'}$ is at least $\frac{2l^*}{\sqrt{i}}$. If this would not be the case, then replacing the edges $\{t_j, h_j\}$ and $\{t_{j'}, h_{j'}\}$ in \hat{t} via a 2-change into $\{t_j, t_{j'}\}$ and $\{h_j, h_{j'}\}$ would result in a shorter tour because the length of each of the former two edges is larger than $\frac{2l^*}{\sqrt{i}}$ by the definition of E_i. This would contradict the local optimality of \hat{t}.

Hence, $|V_u| \geq \sqrt{i}$ heads can be identified in E_i that all have a mutual distance of at least $\frac{2l^*}{\sqrt{i}}$. This implies that the optimal tour on these heads is at least $2l^*$, which gives a contradiction because due to the triangle inequality it is not possible that an optimal tour over all cities is shorter than the optimal tour over a subset of all cities. This proves our claim $|V_u| < \sqrt{i}$. We use this result to derive a contradiction from the assumption $r \geq i$, which settles the proof of the lemma.

Consider the following labeling algorithm. Select an unlabeled tail t_u in E_i and label all tails in E_i that have a distance of at most $\frac{l^*}{\sqrt{i}}$ from t_u (including t_u). Repeat this labeling step until all tails in E_i are labeled. As proved above, fewer than \sqrt{i} tails are labeled in each iteration. As $r \geq i$, this implies that the labeling procedure takes more than \sqrt{i} iterations. Obviously, the tails that are selected in these iterations all have a mutual distance larger than $\frac{l^*}{\sqrt{i}}$. This implies that the optimal tour has length larger than $\sqrt{i} \cdot \frac{l^*}{\sqrt{i}} = l^*$, which results in a contradiction. This proves the lemma. □

Using Lemma 5.3 we can now prove the claimed result.

Theorem 5.5. *The k-change neighborhood function has a performance bound of $4\sqrt{n}$ for* METRIC TSP.

Proof. To prove the theorem it suffices to show that a tour \hat{t} has a performance ratio of at most $4\sqrt{n}$ if it is a locally optimal tour with respect to the 2-change neighborhood function. Lemma 5.3 yields that the ith largest edge in \hat{t} has a length of at most $\frac{2l^*}{\sqrt{i}}$, where l^* is the optimal tour length. Hence, the length of \hat{t} is at most

$$\sum_{i=1}^{n} \frac{2l^*}{\sqrt{i}} = 2l^* \sum_{i=1}^{n} \frac{1}{\sqrt{i}}. \tag{5.1}$$

As $\frac{1}{\sqrt{x}}$ is a decreasing function in $x \in \mathbb{R}$, we have $\frac{1}{\sqrt{i}} \leq \int_{i-1}^{i} \frac{1}{\sqrt{x}} dx$ for any positive integer i. Hence, the right-hand side of (5.1) is at most $2l^* \int_0^n \frac{1}{\sqrt{x}} dx$, which equals $4l^*\sqrt{n}$. □

5.2.2 Performance Ratios of the Move and Swap Neighborhood Functions

In this section we address performance ratios for $Pm||C_{max}$ and $Qm||C_{max}$. For $Pm||C_{max}$ we indicated in Section 2.2 that an obvious representation of a schedule is a partition $P = (A_1, A_2, \ldots, A_m)$ of the n jobs into m subsets, where subset A_i contains the jobs that are processed on machine i with $1 \leq i \leq m$. The same solution representation can be used for $Qm||C_{max}$. For a partition P we define l_i as the amount of time machine i requires for processing all jobs in subset A_i. Hence, l_i is given by $\sum_{j \in A_i} p_j$ for $Pm||C_{max}$ and by $\sum_{j \in A_i} \frac{p_j}{s_i}$ for $Qm||C_{max}$. Recall that s_i gives the speed of machine i. We call l_i the load of machine i. Note that machine i is critical if and only if $l_i = C_{max}$.

In a local search algorithm we can use the move as well as the swap neighborhood function, both of which were introduced in Section 2.2. The move neighborhood function selects a job from a given subset (machine) and assigns it to another subset (machine). The swap neighborhood function extends this neighborhood function by adding solutions to the neighborhood that can be obtained by interchanging two jobs $j \in A_i$ and $j' \in A_{i'}$ from two different subsets, i.e., j is assigned to $A_{i'}$ and j' is assigned to A_i.

Regarding the cost function, we indicated in Section 2.2 that it is better not to simply take the function that returns the makespan of a schedule, but to take the function that returns both the makespan of a schedule and the number of critical machines it contains. A partition P is better than partition P' with respect to this

cost function if it has a smaller makespan or if the makespans of both partitions are equal, but the number of critical machines in P is smaller than in P'.

Lemma 5.4. *The move neighborhood function has a performance bound of* $2 - \frac{2}{m+1}$ *for* $Pm||C_{max}$.

Proof. Let partition $P = (A_1, A_2, \ldots, A_m)$ define a locally but not globally optimal schedule with makespan C_{max}, and let $P^* = (A_1^*, A_2^*, \ldots, A_m^*)$ define a globally optimal schedule with makespan C_{max}^*. To prove the lemma, we show that $C_{max} \leq (2 - \frac{2}{m+1})C_{max}^*$. Without loss of generality, we assume that partition P satisfies $l_1 \geq l_2 \geq \cdots \geq l_m$. Note that $l_1 = C_{max}$. Let job k be a job in A_1 with the smallest processing requirement. As P is locally optimal we obtain $l_i + p_k \geq C_{max}$ for all $2 \leq i \leq m$, which gives

$$mC_{max} \leq l_1 + \sum_{i=2}^{m}(l_i + p_k) = \sum_{j \in J} p_j + (m-1)p_k. \tag{5.2}$$

Obviously, A_1 contains at least two jobs because $C_{max} > C_{max}^*$ and because the maximum processing requirement of any job is a lower bound on C_{max}^*. Hence, we have $p_k \leq \frac{1}{2}C_{max}$. Using this result and $\frac{1}{m}\sum_{j \in J} p_j \leq C_{max}^*$, we can derive from (5.2) that

$$C_{max} \leq C_{max}^* + \frac{m-1}{m}p_k \leq C_{max}^* + \frac{m-1}{2m}C_{max}.$$

This yields $(1 - \frac{m-1}{2m})C_{max} \leq C_{max}^*$, which implies $C_{max} \leq (2 - \frac{2}{m+1})C_{max}^*$. □

As each local optimum for the swap neighborhood function is also a local optimum for the move neighborhood function, Lemma 5.4 implies that the swap neighborhood function also has a performance bound of $2 - \frac{2}{m+1}$. Similarly, if we can prove that this bound is tight for the swap neighborhood function, we obtain that it is also tight for the move neighborhood function. This result follows from the following example.

Example 5.1. Suppose that we have m large jobs with processing requirement Km and $Km + m - 1$ small jobs with unit processing requirement, where $K \geq 1$ is a given positive integer. Figure 5.7(a) gives an optimal schedule with makespan $Km + K + 1$ for this problem instance. Each machine processes one large job and either K (machine 1) or $K + 1$ (the other machines) small jobs. A swap-optimal schedule is depicted in Figure 5.7(b). Machine 1 processes two large jobs, machines $2, 3, \ldots, m-1$ process one large job and one small job, and machine m processes $Km + 1$ small jobs. The makespan of this schedule is $2Km$, which implies that the performance ratio of this locally optimal schedule is $\frac{2Km}{Km+K+1}$. For $K \to \infty$, this ratio approaches $\frac{2m}{m+1} = 2 - \frac{2}{m+1}$. Hence, the performance ratio of the swap neighborhood is at least $2 - \frac{2}{m+1}$. □

Consequently, combining Lemma 5.4 and the result of Example 5.1 gives us the following theorem.

Theorem 5.6. *Both the move and the swap neighborhood function have a performance ratio of* $2 - \frac{2}{m+1}$ *for* $Pm||C_{max}$. □

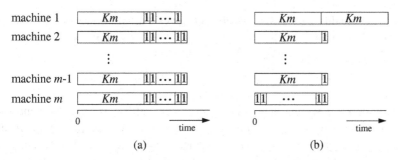

Figure 5.7. (a) Optimal schedule. (b) Locally optimal schedule with respect to swap neighborhood function.

In the remainder of this section we focus on the worst-case performance of the move neighborhood function and the swap neighborhood function for $Qm||C_{max}$.

Lemma 5.5. *The move neighborhood function has a performance bound of* $\frac{1+\sqrt{4m-3}}{2}$ *for* $Qm||C_{max}$.

Proof. Let $P = (A_1, A_2, \ldots, A_m)$ with makespan C_{max} be a locally optimal solution, and let $P^* = (A_1^*, A_2^*, \ldots, A_m^*)$ with makespan C_{max}^* define a globally optimal solution. Without loss of generality we assume $p_1 \geq p_2 \geq \cdots \geq p_n$ and $l_1 \geq l_2 \geq \cdots \geq l_n$, where the l_i's refer to partition P. Furthermore, s_{max} gives the maximum speed of any machine.

Obviously, the total amount of work that has to be processed is given by $\sum_{j=1}^{n} p_j$ and the total amount of work that can be processed in a schedule of length C is given by $\sum_{i=1}^{m} s_i C$. This means that

$$\sum_{j=1}^{n} p_j \leq \sum_{i=1}^{m} s_i C_{max}^* \tag{5.3}$$

has to hold. Furthermore, as the total amount of work $s_{max}C$ that can be processed by the fastest machine on a schedule of length C must be at least the largest processing requirement p_1 of a job, we have

$$p_1 \leq s_{max} C_{max}^*. \tag{5.4}$$

We use (5.3) and (5.4) to prove two upper bounds on C_{max}. Using these two bounds, we next derive the upper bound of $\frac{1+\sqrt{4m-3}}{2}$ on the performance ratio of P.

We have $l_1 = C_{max}$ and, as P is a local optimum, also $l_i + \frac{p_k}{s_i} \geq C_{max}$ for all $2 \leq i \leq m$, where k is a job from machine 1 with minimal processing requirement. Hence, the total amount of work $\sum_{i=1}^{m} s_i C_{max}$ that can be processed in a schedule of length C_{max} satisfies

$$\sum_{i=1}^{m} s_i C_{max} \leq s_1 l_1 + \sum_{i=2}^{m} (s_i l_i + p_k) = \sum_{j=1}^{n} p_j + (m-1)p_k.$$

In the last equality we use the fact that $s_i l_i$ gives the work processed by machine i. Using Equations (5.3) and (5.4) we derive that the right-hand side is at most $\sum_{i=1}^{m} s_i C_{\max}^* + (m-1) s_{\max} \cdot C_{\max}^*$. As a result, we get

$$C_{\max} \leq \left(1 + \frac{(m-1)s_{\max}}{\sum_{i=1}^{m} s_i}\right) C_{\max}^*.$$

We now derive a second upper bound on C_{\max}. Let $2 \leq i \leq m$. As mentioned, we have $C_{\max} = l_1$ and $C_{\max} \leq l_i + \frac{p_k}{s_i}$. Hence, we obtain $s_1 C_{\max} = \sum_{j \in A_1} p_j$ and $s_i C_{\max} \leq (\sum_{j \in A_i} p_j) + p_k$. As job k is assigned to machine 1, i.e., $k \in A_1$, this gives $s_i C_{\max} \leq \sum_{j=1}^{n} p_j$ for each machine $i \geq 1$. By using (5.3) we can now derive

$$C_{\max} \leq \frac{\sum_{j=1}^{n} p_j}{s_{\max}} \leq \frac{\sum_{i=1}^{m} s_i}{s_{\max}} C_{\max}^*,$$

which yields our second upper bound. Combining both upper bounds on C_{\max} gives

$$C_{\max} \leq \min\left(1 + \frac{m-1}{x}, x\right) C_{\max}^*,$$

where $x = \frac{\sum_{i=1}^{m} s_i}{s_{\max}}$. As $1 + \frac{m-1}{x}$ and x are decreasing and increasing in x, respectively, $\min(1 + \frac{m-1}{x}, x)$ is maximal if $1 + \frac{m-1}{x} = x$. Using elementary calculus, this equation can be rewritten as $x = \frac{1 + \sqrt{4m-3}}{2}$, which proves the performance bound. \square

The following example shows that the bound given in Lemma 5.5 is tight. Note that this implies that if m is part of the input, i.e., if we consider $Q||C_{\max}$ instead of $Qm||C_{\max}$, then the move neighborhood function has an unbounded performance ratio. This is in contrast to $P||C_{\max}$, for which, according to Theorem 5.6, the move neighborhood function has a performance ratio of 2.

Example 5.2. Consider the problem instance in which machine 1 has a speed s of $\frac{1 + \sqrt{4m-3}}{2}$ and all other $m-1$ machines have speed one. Note that the definition of s implies that $m = s^2 - s + 1$. Furthermore, assume that we have m jobs. Job 1 has a processing requirement of s and all other jobs have a unit processing requirement.

For this problem instance an optimal schedule with makespan one is obtained by assigning the large job to machine 1 and by assigning one small job to each of the other $m-1$ machines. Now consider the schedule in which the large job is assigned to machine 2 and all other $m-1$ small jobs are assigned to machine 1. Then machine 1 finishes at time $\frac{m-1}{s} = \frac{s^2 - s}{s} = s - 1$, machine 2 at time s, and all other machines at time zero. Hence, the schedule has makespan $C_{\max} = s$. This schedule is locally optimal with respect to the move neighborhood function: machine 2 is the only critical machine and moving the large job from this machine to any other machine does not improve the makespan. Furthermore, it has a performance ratio of $s = \frac{1 + \sqrt{4m-3}}{2}$. \square

We have now proved the following result.

Theorem 5.7. *The move neighborhood function has a performance ratio of* $\frac{1+\sqrt{4m-3}}{2}$ *for* $Qm||C_{\max}$. □

The performance bound of the move neighborhood function given by Lemma 5.5 directly translates to the swap neighborhood function. To determine the quality of this bound consider the problem instance of Example 5.2. We modify this problem instance by adding a single job with unit processing requirement. Then an optimal schedule is obtained from the optimal schedule given in Example 5.2 by adding this job to machine 1. The makespan of this optimal schedule is $1 + \frac{1}{s}$. Moreover, adding this new job to machine 1 in the move optimal schedule presented results in an increase in the finishing time of machine 1 from $s - 1$ to $s - 1 + \frac{1}{s}$. As $s > 1$, the makespan of this schedule is still s. Furthermore, it is swap optimal, as can be verified. Hence, the performance ratio of the swap neighborhood function is at least $\frac{s}{1+1/s}$ with $s = \frac{1+\sqrt{4m-3}}{2}$. Clearly, $s - \frac{s}{1+1/s} = \frac{1}{1+1/s}$, which is the difference between the performance bound s given in Lemma 5.5 and this lower bound on the performance ratio, is an increasing function in s, and it approaches 1 for $s \to \infty$. This gives the following result.

Theorem 5.8. *For* $Qm||C_{\max}$, *the swap neighborhood function has a performance ratio R that satisfies*

$$\frac{1+\sqrt{4m-3}}{2} - 1 \leq R \leq \frac{1+\sqrt{4m-3}}{2}.$$

□

5.3 Non-Approximability Results

Many of the interesting combinatorial optimization problems belong to the complexity class NPO. For a definition of this natural optimization counterpart of the more well-known complexity class NP of decision problems, we refer to Appendix B. Let Π be an arbitrary problem in NPO, and let N be a neighborhood function that we want to use for it in a local search algorithm. Nearly all neighborhood functions used in practice have the property that they are polynomially searchable. By this we mean the following.

Definition 5.3. A neighborhood function is said to be *polynomially searchable* if in polynomial time we can decide whether a solution is locally optimal and, if not, construct a better neighboring solution. □

For iterative improvement, being polynomially searchable implies that an implementation exists for which each iteration runs in polynomial time. Note that this is obviously the case if the neighborhood is of polynomial size.

Suppose that finding an R-approximate solution for Π is NP-hard for a given $R \geq 1$. Provided that P \neq NP, this means that iterative improvement using neighborhood function N does not find a solution with performance ratio R in polynomial time. As it may take iterative improvement an exponential number of steps to reach a local optimum, this does not directly imply that N has a performance ratio larger

than R (see also the beginning of Section 5.2.1). The following theorem states that the claim is nevertheless true. Observe that, by taking $R = 1$, the theorem yields that an NP-hard problem will not admit a polynomially searchable exact neighborhood function.

Theorem 5.9. *Let $\Pi \in NPO$ and let N be a neighborhood function for Π that is polynomially searchable. If the problem of finding an R-approximate solution for Π is NP-hard for a given $R \geq 1$, then N does not have a performance bound of R, provided that $NP \neq co\text{-}NP$.*

Proof. We prove the result by contradiction. Suppose that approximating Π within a factor R is NP-hard and that N has a performance bound of R. We show that for any decision problem $\Pi_D \in NP$ the complementary problem Π_D^c obtained from Π_D by reversing the answers is also in NP. This yields a contradiction as it implies $NP = co\text{-}NP$.

To prove $\Pi_D^c \in NP$, we need to show that each yes-instance of Π_D^c has a certificate that can be checked in polynomial time for validity, where the size of the certificate is polynomially bounded. As approximating Π within a factor R is NP-hard, a polynomial-time algorithm \mathcal{A} exists that decides Π_D, where \mathcal{A} may use an oracle that returns an R-approximate solution for any problem instance of Π. Hence, if for a problem instance I of Π_D we are given a sequence consisting of solutions that may be returned successively by the oracle during the execution of \mathcal{A} on I, then we can determine in polynomial time whether I is a yes-instance or a no-instance of Π_D. This implies that as a certificate for yes-instances of Π_D^c we can use a sequence consisting of solutions that are locally optimal for the problem instances that are successively given to the oracle, i.e., the ith solution in the sequence is a local optimum for the ith problem instance given to the oracle. The certificate, which is of polynomial size, can be checked for validity in polynomial time by substituting each call of the oracle in \mathcal{A} by a polynomial-time procedure that checks whether the corresponding solution in the certificate is a local optimum for the problem instance given to the oracle. By assumption this polynomial-time procedure exists. □

From the above theorem and Theorem 2.1 it follows that, if $NP \neq co\text{-}NP$, then the performance ratio of the k-change neighborhood function cannot be bounded by $2^{p(n)}$ for some fixed polynomial p. Note that this result is weaker than the result of Theorem 5.4. Theorem 5.4 states that the performance ratio cannot even be bounded by 2^{2^n}, for instance, and that it also holds in the unlikely case that $NP = co\text{-}NP$.

5.3.1 Exact Neighborhood Function for TSP: An NP-Completeness Proof

Theorems 2.1 and 5.9 yield that, unless $NP = co\text{-}NP$, no polynomially searchable exact neighborhood function exists for METRIC TSP. In this section we give an alternative proof of this result. More precisely, we prove that deciding whether a tour is suboptimal or, equivalently, not locally optimal with respect to an exact neighborhood function is NP-complete. This implies the claimed result, where the condition $NP \neq co\text{-}NP$ is replaced by the weaker condition $P \neq NP$. The main reason

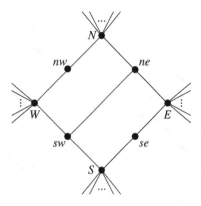

Figure 5.8. A diamond subgraph.

for giving the proof despite the fact that we already have the result is that it is elegant and serves as a good warm-up for the next chapter.

An outline of the proof is as follows. It is well known that HAMILTONIAN CYCLE (HC), which is the problem of deciding for a graph whether it contains a Hamiltonian cycle, is NP-complete. We first prove that the problem does not become easier if in addition to the graph we are also given a Hamiltonian path. Remember that a Hamiltonian cycle is a Hamiltonian path in which we return to the starting node. Hence, we prove that the following problem is NP-complete.

Definition 5.4 [RESTRICTED HAMILTONIAN CYCLE (RHC)]. Given are a graph G and a Hamiltonian path in G. Is there a Hamiltonian cycle in G? □

Using this result, we next show that METRIC TSP SUBOPTIMALITY is also NP-complete, which was our goal.

Definition 5.5 [METRIC TSP SUBOPTIMALITY]. Given are an instance I of METRIC TSP and a tour τ for I. Is τ suboptimal? □

Lemma 5.6. RHC *is NP-complete.*

Proof. Proving RHC \in NP is trivial. Hence, to prove the lemma it now suffices to show that RHC is polynomially reducible from HC. Let $G = (V, E)$ be an arbitrary graph defining a problem instance of HC. The construction of a corresponding instance of RHC is based on the special-purpose subgraph depicted in Figure 5.8. This subgraph, which consists of eight nodes, is called a diamond.

In a diamond we identify four special nodes, called N (north), E (east), S (south), and W (west). We will use diamonds in such a way that if a graph G' contains a diamond D, then only these special nodes may be connected to other nodes in the graph. We now claim that if a graph G' contains a Hamiltonian cycle c, then for each diamond D in G' we have that its nodes are visited by c either in the way depicted in Figure 5.9(a) or in the way depicted in Figure 5.9(b). The former case is referred to as north-south mode and the latter as east-west mode.

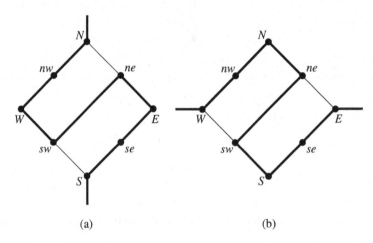

Figure 5.9. (a) North-south mode. (b) East-west mode.

We now focus on proving this claim. Suppose that Hamiltonian cycle c enters diamond D in node N. We show that following any other path than the one depicted in Figure 5.9(a) results in a contradiction. If node N is not succeeded by node nw (north-west), but by ne, then the only way for c to visit nw is via node W. However, this is not possible because, being in nw, we can neither go to N nor return to W as this would imply visiting N or W twice. Hence, N is indeed succeeded by nw in c. Obviously, the only way to proceed from nw is to go to W. Once at node W, we cannot leave D because then the only way for c to visit sw and ne is by skipping se. From node sw we have to go to ne because if we chose S this would make it impossible to visit ne as well as se. From ne we can then only go to E and, to prevent skipping node se, we conclude the traversal of D by visiting se and S, successively. By a similar reasoning, it can be verified that if c enters a diamond D via the south, then D is also traversed as depicted in Figure 5.9(a) and if c enters D via the east or the west, then D is traversed as depicted in Figure 5.9(b). This proves our claim.

We now construct for our arbitrary graph $G = (V, E)$ specifying the arbitrary HC instance the following corresponding graph $G' = (V', E')$. Let $V = \{1, 2, \ldots, n\}$. For each node $i \in V$, we introduce a diamond D_i with special nodes N_i, E_i, S_i, and W_i. Except for the last diamond D_n, we connect the south node of each diamond D_i with the north node of the next diamond D_{i+1}, i.e., we add the edges $\{S_i, N_{i+1}\}$ for all $1 \le i < n$ to E'. Finally, we introduce for each edge $\{i, j\} \in E$ the two edges $\{W_i, E_j\}$ and $\{W_j, E_i\}$.

Obviously, the path that visits the diamonds in increasing order and traverses each diamond from north to south in the way depicted in Figure 5.9(a) gives us a Hamiltonian path. Hence, to prove that we presented a valid polynomial-time reduction, it now suffices to show that G has a Hamiltonian cycle if and only if G' has one. We start with the 'only-if' part. Hence, suppose that G has a Hamiltonian

cycle c. Then visiting the diamonds in G' in the order specified by c and the nodes inside a diamond in east-west mode yields a Hamiltonian cycle for G'. Next, assume that G' has a Hamiltonian cycle c. If c traverses one diamond in north-south mode, then it must traverse all diamonds in north-south mode as north and south nodes are not connected to east and west nodes. However, as only $n-1$ north-south edges exist, no Hamiltonian cycle can exist that traverses each diamond in north-south mode. Hence, we can conclude that c must traverse each diamond in east-west mode. The order in which these diamonds are visited defines a Hamiltonian cycle for G. □

Theorem 5.10. METRIC TSP SUBOPTIMALITY *is NP-complete.*

Proof. Again, proving membership in NP is a trivial task. We now show that RHC is polynomially reducible to METRIC TSP SUBOPTIMALITY. Let $G = (V, E)$ with Hamiltonian path $p = (v_1, v_2, \ldots, v_n)$ define an arbitrary problem instance I of RHC. If $\{v_1, v_n\} \in E$, then we can directly conclude that I is a yes-instance of RHC. Hence, without loss of generality, we assume that $\{v_1, v_n\} \notin E$. We transform problem instance I into the following problem instance of METRIC TSP SUBOPTIMALITY. The set of cities is given by $C = V = \{1, 2, \ldots, n\}$ and the distance d_{ij} between cities i to j is defined by 1 if $\{i, j\} \in E$ and by 2 if $\{i, j\} \notin E$. Hamiltonian path p gives us a tour τ with cost $n+1$ because $\{v_i, v_{i+1}\} \in E$ with $1 \le i < n$ and $\{v_1, v_n\} \notin E$. Now, G has a Hamiltonian cycle if and only if tour τ is suboptimal. To see this, suppose that τ is suboptimal. A tour τ' then exists with length n. This tour defines a Hamiltonian cycle in G. Conversely, if G contains a Hamiltonian cycle, then this cycle determines a tour with length n, which implies that τ is suboptimal. This proves the theorem. □

Corollary 5.1. *Provided that $P \ne NP$, no polynomially searchable exact neighborhood function exists for* METRIC TSP. □

5.4 *From Neighborhood Function to Polynomial-Time Algorithm

Rewriting Theorem 5.9 gives that if a polynomially searchable neighborhood function N with performance bound R exists for some given combinatorial optimization problem $\Pi \in$ NPO, then the problem of finding an R-approximate solution for Π is not NP-hard, provided that NP \ne co-NP. One question that arises is whether this result can be strengthened to the extent that the existence of such an N implies that the problem of finding an R-approximate solution for Π is easy, i.e., polynomially solvable. In this section we prove for a special type of combinatorial optimization problem that the answer is affirmative for $R = 1$ and that it is affirmative up to some arbitrary small error $\varepsilon > 0$ for $R > 1$. In other words, we show for a special type of combinatorial optimization problem how we can derive from a polynomially searchable exact neighborhood function a polynomial-time algorithm that solves the problem and how we can derive from a polynomially searchable neighborhood function with performance bound R a polynomial-time approximation algorithm with a performance bound of $R + \varepsilon$ for any precision $\varepsilon > 0$. Besides being polynomial in

the input size, the $(R+\varepsilon)$-approximation algorithm is also polynomial in $\frac{1}{\varepsilon}$. After introducing some terminology in Section 5.4.1 we prove the former result in Section 5.4.2 and the latter result in Section 5.4.3.

5.4.1 Preliminaries

For many combinatorial optimization problems it holds that a solution can be viewed as a subset of a ground set E and that the cost of a solution $s \subseteq E$ can be written as $\sum_{e \in s} c(e)$, where $c(e)$ is the cost of element $e \in E$. This is, for instance, the case for TSP: define E as the set of all pairs (i,j) of cities and $c(i,j)$ as the distance d_{ij} from city i to city j. Other examples are STG and UGP. We call such problems linear combinatorial optimization problems.

Definition 5.6. An *instance* of a linear combinatorial optimization problem is a pair (S,c), where each solution in the solution space $S \subseteq 2^E$ is a subset of a finite ground set $E = \{1,2,\ldots,n\}$ and where $c : E \to \mathbb{Q}^+$ assigns a cost to each element $e \in E$. The cost of a solution $s \in S$ is defined by $f(s) = \sum_{e \in s} c(e)$. □

Definition 5.7. A *linear combinatorial optimization problem* Π is specified by a set of problem instances as defined in Definition 5.6, and the goal is to minimize for a problem instance the associated cost function f. Hence, for a problem instance (S,c) we have to find a solution s^* that satisfies $f(s^*) \leq f(s)$ for all $s \in S$. □

Definition 5.8. A linear combinatorial optimization problem Π is called *closed under scaling* whenever for each problem instance (S,c) of Π it holds that if we change function $c : E \to \mathbb{Q}^+$ to some other cost function $c' : E \to \mathbb{Q}^+$, then the problem instance remains in Π. □

In this section the focus is on linear combinatorial optimization problems Π that satisfy the following two properties.

- Π is closed under scaling.
- For each problem instance of Π we can derive in polynomial time a feasible solution.

It can easily be verified that the above-mentioned linear combinatorial optimization problems TSP, STG, and UGP all satisfy these two properties. However, this is not true, for instance, for METRIC TSP because this problem is not closed under scaling.

If a problem is closed under scaling, then this need not necessarily mean that we can change the cost function of a problem instance in polynomial time. The reason for this is that an encoding scheme may be applied that does not give $c(e)$ explicitly for each $e \in E$. To avoid this problem we focus on linear combinatorial optimization problems that are encoded by a cost-explicit encoding scheme, where a cost-explicit encoding scheme is defined as follows.

Definition 5.9. An encoding scheme of a linear combinatorial optimization problem is called *cost explicit* if the encoding $e_I \in \{0,1\}^*$ of a problem instance $I = (S,c)$ with underlying ground set E is given by $e_I = e_I^{(1)} e_I^{(2)}$, where

- $e_I^{(1)}$ is an explicit encoding of the function $c : E \rightarrow \mathbb{Q}^+$, which means that it contains the encoding of $c(e)$ for each $e \in E$, and

- $e_I^{(2)}$ is an encoding, either implicit or explicit, of the solution space S.

Furthermore, $e_I^{(1)}$ only depends on S via its ground set E and $e_I^{(2)}$ is independent of c. □

5.4.2 Exact Neighborhood Functions

Let Π be a linear combinatorial optimization problem from NPO that satisfies the two properties given above, and let a cost-explicit encoding scheme be used for Π. Furthermore, let N be an exact polynomially searchable neighborhood function for the problem. In this section we present a polynomial-time algorithm, called $\mathcal{A}_{\text{OPT}}(N)$, that solves Π.

Let $I = (S, c)$ be a problem instance of Π, where, without loss of generality, we assume that c is a function to the natural numbers instead of the rational numbers. $\mathcal{A}_{\text{OPT}}(N)$ solves I in $K = \lceil \log_2(c_{\max} + 1) \rceil$ phases, where c_{\max} is the maximum cost $c(e)$ of any element $e \in E$. In phase k an optimal solution is derived for the problem instance $I_k = (S, c_k)$ that is obtained from I by defining the cost $c_k(e)$ of an element $e \in E$ as the number represented by the k leading bits in the binary representation of $c(e)$. By adding leading zeros, the binary representation is made equally long for all $e \in E$, i.e., each binary representation consists of exactly K bits. We implement a phase by applying iterative improvement with the exact neighborhood function N. In the first phase an arbitrary solution may be chosen as the starting solution, and in phase k with $k \geq 2$ the solution derived in phase $k - 1$ is chosen as the starting solution. Figure 5.10 gives the algorithm in pseudo-code, where f_k denotes the cost function related to problem instance I_k.

Theorem 5.11. *Let $\Pi \in NPO$ be a linear combinatorial optimization problem that is closed under scaling and for which we can derive a feasible solution in polynomial time. Furthermore, let N be a polynomially searchable exact neighborhood function for Π. Then the algorithm $\mathcal{A}_{\text{OPT}}(N)$ given in Figure 5.10 solves Π in polynomial time in the case that we use a cost-explicit encoding scheme for Π.*

Proof. By construction, the number of phases is bounded by $K = \lceil \log_2(c_{\max} + 1) \rceil$. Furthermore, because N is polynomially searchable, the individual iterations of the iterative improvement algorithm that makes up a single phase of $\mathcal{A}_{\text{OPT}}(N)$ can be implemented to run in polynomial time. Hence, to prove the theorem, we only have to show that in an arbitrary phase k the iterative improvement algorithm executed terminates within a polynomial number of iterations.

Let s_i be the solution derived in phase i of $\mathcal{A}_{\text{OPT}}(N)$, and let s_0 be the starting solution of the first phase. Hence, in phase k iterative improvement starts with s_{k-1} and ends with s_k. Because the cost of the solution derived by iterative improvement decreases in each iteration, it suffices to prove that $f_k(s_{k-1}) - f_k(s_k)$ is polynomially bounded to show that in phase k iterative improvement terminates within a polynomial number of iterations. By defining $c^{(i)}(e)$ with $e \in E$ as the ith bit in the binary

algorithm $\mathcal{A}_{\mathrm{OPT}}(N)$

begin
 $s :=$ some initial solution;
 $K := \lceil \log_2(c_{\max} + 1) \rceil$;
 for $k := 1$ **to** K **do**
 begin
 for all $e \in E$ **do** $c_k(e) := \lfloor c(e) \cdot 2^{k-K} \rfloor$;
 repeat
 generate an $s' \in N(s)$;
 if $f_k(s') < f_k(s)$ **then** $s := s'$;
 until $f_k(s') \geq f_k(s)$ for all $s' \in N(s)$;
 end
end;

Figure 5.10. Algorithm $\mathcal{A}_{\mathrm{OPT}}(N)$ for a linear combinatorial optimization problem Π and an accompanying neighborhood function N. Π is closed under scaling and (S, c) denotes an arbitrary problem instance of Π that is given as input to the algorithm. Cost function f_k relates to problem instance (S, c_k).

representation of $c(e)$, we obtain that for any solution s the cost $f_k(s)$ satisfies

$$f_k(s) = \begin{cases} \sum_{e \in s} c^{(1)}(e) & \text{if} \quad k = 1 \\ 2 f_{k-1}(s) + \sum_{e \in s} c^{(k)}(e) & \text{if} \quad k \geq 2. \end{cases}$$

Furthermore, we have $\sum_{e \in s} c^{(k)}(e) \leq n$, where n denotes the number of elements in E. Using these two observations and the optimality of s_{k-1} with respect to cost function f_{k-1}, we can derive for $k \geq 2$

$$f_k(s_{k-1}) - f_k(s_k) = 2\left(f_{k-1}(s_{k-1}) - f_{k-1}(s_k)\right) + \sum_{e \in s_{k-1}} c^{(k)}(e) - \sum_{e \in s_k} c^{(k)}(e) \leq n$$

and for $k = 1$

$$f_k(s_{k-1}) - f_k(s_k) = \sum_{e \in s_{k-1}} c^{(k)}(e) - \sum_{e \in s_k} c^{(k)}(e) \leq n.$$

This proves the theorem. \Box

5.4.3 Neighborhood Functions with Performance Bound

Theorem 5.11 yields that if a polynomially searchable neighborhood function with performance bound $R = 1$ exists for a given linear combinatorial optimization problem Π that satisfies some properties, then the problem of finding an R-approximate solution for Π is polynomially solvable. In this section we show that for $R > 1$ this result holds up to some arbitrary small error $\varepsilon > 0$. More precisely, we proceed as follows.

Let Π be a linear combinatorial optimization problem from NPO that satisfies the same two properties as those given in Theorem 5.11 and for which we use a cost-explicit encoding scheme. Furthermore, let N be a polynomially searchable neighborhood function for Π with a performance bound of R. We now present an algorithm for which we will prove that for any precision $\varepsilon > 0$ it (i) has a running time that is polynomially bounded in both the input size and $\frac{1}{\varepsilon}$ and (ii) has a performance bound of $R + \varepsilon$.

Let $I = (S,c)$ be a problem instance of Π, and suppose that we perform iterative improvement on I, where we take an arbitrary $s \in S$ as a starting solution. By assumption, s can be derived in polynomial time. We turn iterative improvement into a polynomial-time algorithm in the following way: we replace the function c in I by c', such that in each iteration the cost of the solution derived by iterative improvement decreases by at least q and $f'(s)/q$ is polynomially bounded, where $f'(s)$ is the cost of solution s with respect to the new function $c' : E \rightarrow \mathbb{Q}^+$. The former property is achieved by defining for any $e \in E$

$$c'(e) = \left\lceil \frac{c(e)}{q} \right\rceil q.$$

The latter property is achieved by defining q as

$$q = \frac{f(s)\varepsilon}{2nR(R+\varepsilon)}$$

since we then have

$$\frac{f'(s)}{q} \leq \frac{f(s) + nq}{q} = 2nR\left(\frac{R}{\varepsilon} + 1\right) + n,$$

which is polynomially bounded in $|I|$ and $\frac{1}{\varepsilon}$.

Using the polynomial-time algorithm described, we can now construct a polynomial-time $(R+\varepsilon)$-approximation algorithm $\mathcal{A}_\varepsilon(N)$ for Π. Algorithm $\mathcal{A}_\varepsilon(N)$ proceeds in phases. In each phase the algorithm described is executed, where the final solution of the previous phase is taken as a starting solution. If the algorithm arrives at a solution for which the cost is at most half the cost of the solution with which the phase started, then the algorithm is interrupted and the next phase is initiated. If the algorithm terminates, i.e., if it arrives at a local optimum, then the overall algorithm $\mathcal{A}_\varepsilon(N)$ also terminates. In Figure 5.11 algorithm $\mathcal{A}_\varepsilon(N)$ is given in pseudo-code.

Lemma 5.7. *Let $\Pi \in$ NPO be a linear combinatorial optimization problem that is closed under scaling and for which we can derive a feasible solution in polynomial time. Furthermore, let N be a polynomially searchable neighborhood function for Π. Then the algorithm $\mathcal{A}_\varepsilon(N)$ given in Figure 5.11 runs in polynomial time in the case that we use a cost-explicit encoding scheme for Π.*

Proof. The number of phases executed by $\mathcal{A}_\varepsilon(N)$ is bounded by $\log(nc_{\max})$, where c_{\max} is the maximum cost $c(e)$ of any element $e \in E$. As a cost-explicit encoding scheme is assumed, this implies that the number of executed phases is polynomially

algorithm $\mathcal{A}_\varepsilon(N)$

begin
 $s :=$ some initial solution;
 $stop :=$ FALSE;
 while $\neg stop$ **do**
 begin
 $s_{\text{init}} := s$;
 $q := f(s_{\text{init}})\varepsilon/(2nR(R+\varepsilon))$;
 for all $e \in E$ **do** $c'(e) := \left\lceil \frac{c(e)}{q} \right\rceil q$;
 repeat
 if $f'(s') \geq f'(s)$ for all $s' \in N(s)$ **then** $stop :=$ TRUE;
 else
 begin
 generate an $s' \in N(s)$;
 if $f'(s') < f'(s)$ **then** $s := s'$;
 end
 until $stop =$ TRUE or $f(s) \leq f(s_{\text{init}})/2$;
 end
end;

Figure 5.11. Algorithm $\mathcal{A}_\varepsilon(N)$ for a linear combinatorial optimization problem Π that is closed under scaling, an accompanying neighborhood function N, and an arbitrary precision $\varepsilon > 0$. (S,c) denotes an arbitrary problem instance of Π that is given as input to the algorithm. Cost function f relates to problem instance (S,c), while cost function f' relates to problem instance (S,c').

bounded. Combined with the above-derived polynomial running time of each individual phase, this proves the polynomial-time running time of $\mathcal{A}_\varepsilon(N)$. □

The next lemma focuses on the quality of the solutions returned by $\mathcal{A}_\varepsilon(N)$.

Lemma 5.8. *Let Π be a linear combinatorial optimization problem that is closed under scaling, and let N be a neighborhood function for Π with performance bound R. Then the algorithm $\mathcal{A}_\varepsilon(N)$ given in Figure 5.11 is an $(R+\varepsilon)$-approximation algorithm.*

Proof. Let s_{init}, q, c', and f' denote the corresponding values at the moment $\mathcal{A}_\varepsilon(N)$ terminates, let s be the solution returned by the algorithm, and let s^* be an optimal solution to the problem instance $I = (S,c)$, which is the problem instance that is given as input to the algorithm.

According to the definition of function c', we have $f(s) \leq f'(s)$. Furthermore, as s is locally optimal for problem instance $I' = (S,c')$ (but not necessarily for I) and as the neighborhood function has a performance bound of R, by assumption, we

have $f'(s) \leq R \cdot f'(s^*)$. Using this, we can derive

$$f(s) \leq R \cdot f'(s^*) = R \sum_{e \in s^*} \left\lceil \frac{c(e)}{q} \right\rceil q \leq R \sum_{e \in s^*} (c(e) + q) \leq R \cdot f(s^*) + Rnq. \quad (5.5)$$

Moreover, $f(s_{\text{init}})/2 < f(s)$ holds because otherwise the last phase of the algorithm would have been interrupted just before setting the variable $stop$ to true. From this and the definition of q it now follows that $q < f(s)\frac{\varepsilon}{nR(R+\varepsilon)}$. Combined with (5.5) this gives $f(s) \leq R \cdot f(s^*) + f(s)\frac{\varepsilon}{R+\varepsilon}$, which can be rewritten as $f(s) \leq (R+\varepsilon)f(s^*)$.
□

Theorem 5.12. *Let $\Pi \in NPO$ be a linear combinatorial optimization problem that is closed under scaling and for which we can derive a feasible solution in polynomial time. Furthermore, let N be a polynomially searchable neighborhood function for Π with performance bound R. Then the algorithm $\mathcal{A}_\varepsilon(N)$ given in Figure 5.11 is a polynomial-time $(R + \varepsilon)$-approximation algorithm in the case that we use a cost-explicit encoding scheme for Π.*

Proof. The theorem follows directly from Lemmas 5.7 and 5.8. □

5.5 Bibliographical Notes

In Section 5.1 we mentioned bubble sort as an example of an iterative improvement algorithm with an exact neighborhood function. Knuth [1998] presents an extensive study on this algorithm and other sorting algorithms. The proposed exact neighborhood for MINIMUM SPANNING TREE is due to Papadimitriou & Steiglitz [1982].

Section 5.1.1 deals with EUCLIDEAN TSP with a bounded number n_{in} of interior cities. Deineko, Hoffmann, Okamoto & Woeginger [2004] prove that the problem is polynomially solvable if $n_{\text{in}} = \mathcal{O}(\log n)$.

Section 5.2.1 is based on the paper of Chandra, Karloff & Tovey [1999]. Besides proving the upper bound $4\sqrt{n}$ on the performance ratio of the k-change neighborhood function when applied to METRIC TSP, they also prove that, for infinitely many values of n, a lower bound is given by $\frac{1}{4}\sqrt{n}$ for $k = 2$ and by $\frac{1}{4}n^{\frac{1}{2k}}$ for $k \geq 3$. Moreover, they prove that k-change has a performance bound of $\mathcal{O}(\log n)$ for EUCLIDEAN TSP in the m-dimensional space. For standard EUCLIDEAN TSP, i.e., for $m = 2$, they derive a lower bound of $c\frac{\log n}{\log \log n}$ on the performance ratio of 2-change for some $c > 0$. We proved Theorem 5.4 by constructing a problem instance of SYMMETRIC TSP for which k-change induces an arbitrarily large performance ratio. An alternative problem instance is given by Papadimitriou & Steiglitz [1978]. Besides having an arbitrarily large performance ratio, their problem instance also has the unfavorable property that the number of locally optimal tours with such a performance ratio is exponential in k. For the special case of METRIC TSP in which the distance between two cities is either 1 or 2, Khanna, Motwani, Sudan & Vazirani [1998] prove that 2-change has a performance bound of $\frac{3}{2}$; see Exercise 3. They claim that this bound is asymptotically tight.

Finn & Horowitz [1979] prove Lemma 5.4, which states that the move neighborhood function for MULTIPROCESSOR SCHEDULING has a performance bound

of $2 - \frac{2}{m+1}$. The proof we presented is taken from Vredeveld [2002]. By giving a problem instance with a performance ratio of $2 - \frac{2}{m+1}$, Langston [1982] proves that this bound is tight for the move neighborhood function and Vredeveld [2002] proves that it is also tight for the swap neighborhood function. The problem instance given in Example 5.1 is similar to the one presented by Langston [1982].

The proof of Lemma 5.5 is due to Cho & Sahni [1980]. However, they did not present it in the context of the move neighborhood function for $Qm||C_{max}$, but to prove that $\frac{1+\sqrt{4m-3}}{2}$ is a performance bound for the list scheduling algorithm when applied to $Qm||C_{max}$. Example 5.2 is presented by Vredeveld [2002]. Vredeveld [2002] also derives lower bounds on the performance ratio of the swap and move neighborhood function when applied to $Rm||C_{max}$; see Exercise 4.

Schuurman & Vredeveld [2001] and Vredeveld [2002] propose the push neighborhood function as an alternative to the move and swap neighborhood functions. A push corresponds to a sequence of moves in the move neighborhood function. They also analyze the performance ratio of this neighborhood function for $Pm||C_{max}$ and $Qm||C_{max}$.

We now mention other (classes of) problems for which performance guarantees for local search have been derived. Performance bounds for satisfiability problems are proved by Alimonti [1995], Alimonti [1997], Halldórsson & Lau [1997], Hansen & Jaumard [1990], Hirsch [2003], and Khanna et al. [1998]. Furthermore, Dantsin et al. [2002] gives an exact neighborhood function for kSAT. In this satisfiability problem we are asked to satisfy all clauses or to conclude that this is not possible, where the number of literals in each clause is at most k. Bylka, Idzik & Tuza [1999] and Feigle, Karpinski & Langberg [2002] derive performance guarantees of local optima for MAX-CUT, Boykov, Veksler & Zabih [1998] for MULTIWAY MINIMUM CUT, and Boykov, Veksler & Zabih [2001] and Gupta & Tardos [2000] for the METRIC LABELING PROBLEM, which is a generalization of MULTIWAY MINIMUM CUT. For several facility location problems Arya et al. [2004], Charikar & Guha [1999], Hassin, Levin & Morad [2003], Kanungo et al. [2002], and Korupolu, Plaxton & Rajaraman [2000] derive approximation guarantees for local search. For several spanning tree problems this is done by Brüggemann, Monnot & Woeginger [2003], Lu & Ravi [1992], and Raghavachari [1997]. The quality of local optima for the QUADRATIC ASSIGNMENT PROBLEM is studied by Angel & Zissimopoulos [1998]. Finally, we refer to the performance bounds of local optima that are derived for INDEPENDENT SET by Khanna, Motwani, Sudan & Vazirani [1998] and for the weighted and unweighted variants of k-SET PACKING by Arkin & Hassin [1998], Bafna, Narayanan & Ravi [1996], Berman [2000], Chandra & Halldórsson [2001], Halldórsson [1995], and Hurkens & Schrijver [1989].

Yannakakis [1997] 'proved' a slightly stronger variant of Theorem 5.9 in the sense that if approximating a problem within a factor R is *strongly* NP-hard, then the condition NP \neq co-NP can be replaced by the weaker condition P \neq NP. However, as the 'proof' contains an illegal step, we resorted to the variant presented.

Papadimitriou & Steiglitz [1977] derived the result that, unless $P = NP$, no polynomially searchable exact neighborhood function exists for SYMMETRIC TSP that has a constant performance ratio. The proof is similar to the one given in Section 5.3.1, which is based on Papadimitriou & Steiglitz [1982]; see also Exercise 6. Armstrong & Jacobson [2003] prove that, unless $P = NP$, no polynomially searchable exact neighborhood function exists for the unweighted version of MAX-2SAT and for VERTEX COVER. Like Papadimitriou and Steiglitz, they use the theory of NP-completeness to prove this result. Exercise 7 is taken from the paper of Armstrong & Jacobson [2003].

Theorem 5.11 has been proved by Schulz, Weismantel & Ziegler [1995] and Theorem 5.12 by Orlin, Punnen & Schulz [2004]. The authors show that both algorithm $\mathcal{A}_{OPT}(N)$ described in Figure 5.10 and algorithm $\mathcal{A}_{\varepsilon}(N)$ described in Figure 5.11 can be turned into strongly polynomial algorithms, i.e., algorithms with a running time that is independent of the encoding length of the numbers that appear in the problem.

Finally, we note that Exercise 5 of this chapter is due to Punnen, Margot & Kabadi [2003].

5.6 Exercises

1. In Section 5.1 we indicated that sorting a multiset of n numbers a_1, a_2, \ldots, a_n non-increasingly corresponds to finding a permutation π of $\{1, 2, \ldots, n\}$ that minimizes the cost function

$$f(\pi) = \sum_{i=1}^{n} i \cdot a_{\pi(i)}.$$

Furthermore, we showed that we can find a permutation that minimizes $f(\pi)$ by performing iterative improvement based on the exact neighborhood function in which a permutation π' is a neighbor of permutation π if and only if it can be derived by swapping two adjacent numbers. Consider the two alternative cost functions

$$g(\pi) = \sum_{i=2}^{n} s(a_{\pi(i)} - a_{\pi(i-1)})$$

and

$$h(\pi) = \sum_{i=2}^{n} \sum_{j=1}^{i-1} s(a_{\pi(i)} - a_{\pi(j)}),$$

where $s(x) = 1$ if $x > 0$ and $s(x) = 0$ if $x \leq 0$.

a) Prove that the cost functions g and h are indeed alternatives to cost function f, i.e., prove that if π is globally optimal with respect to minimizing g or f, then it defines a non-increasing ordering of the numbers.

b) Is the neighborhood function discussed exact for cost function g? And for h? If so give a proof, if not give a counterexample.

2. Consider the following problem.

 Definition 5.10 [MAXIMUM CLAUSE-WEIGHTED SATISFIABILITY]. Given are a set U of binary variables, a set C of clauses over U, and a weight $w(c) \in \mathbb{N}^+$ for each clause $c \in C$. A clause is satisfied by a given truth assignment if and only if at least one of its literals is true. Find a truth assignment that maximizes the sum of the weights of the clauses that are satisfied. □

 A possible neighborhood function for this problem is flip, in which a truth assignment t' is a neighbor of truth assignment t if and only if t' can be obtained from t by changing the value of one variable. Prove that the flip neighborhood function has a performance bound of two.

3. Let $(1,2)$-TSP be the special case of SYMMETRIC TSP in which the distance between any two cities is either 1 or 2.

 a) Prove or disprove that $(1,2)$-TSP is also a special case of METRIC TSP.

 Let $\hat{\tau}$ be a locally optimal tour with respect to the 2-change neighborhood function and let τ^* be a globally optimal tour. We define $U(\tau)$ as the set of edges in tour τ that have a length of 1.
 With each edge $e = (\hat{\tau}(i), \hat{\tau}(j)) \in U(\tau^*)$ we associate an edge $e' \in U(\hat{\tau})$ in the following way. If e is also an edge from $\hat{\tau}$, which means that $j = i+1$, then $e' = e$. Otherwise, e' is an edge from $\{(\hat{\tau}(i), \hat{\tau}(i+1)), (\hat{\tau}(j), \hat{\tau}(j+1))\}$ that has a length of 1.

 b) Prove that edges $(\hat{\tau}(i), \hat{\tau}(i+1))$ and $(\hat{\tau}(j), \hat{\tau}(j+1))$ cannot both have a length of 2.

 Let V be the set of edges from $\hat{\tau}$ that are mapped to an edge from $U(\tau^*)$.

 c) Prove that $|V| \geq |U(\tau^*)|/2$.

 d) Prove that the 2-change neighborhood function has a performance bound of $\frac{3}{2}$ for $(1,2)$-TSP.

4. Consider $Rm||C_{max}$ and let $p_{max} = \max_{i,j} p_{ij}$ denote the maximum processing requirement for a given problem instance and let C_{max}^* denote the optimal makespan.

 a) Show that the move neighborhood function has a performance ratio of at least p_{max}/C_{max}^*.

 b) Show that the swap neighborhood function has a performance ratio of at least $(p_{max} - 1)/C_{max}^*$.

 c) Show that for $m = 2$ the swap neighborhood function has a performance ratio of at least $n - 1$.

5. Consider SYMMETRIC TSP. For a given problem instance $I = (S, f)$ we define avg_I as the average length of all tours, i.e.,

$$avg_I = \sum_{\tau \in S} \frac{f(\tau)}{|S|}.$$

In this exercise it is our goal to prove that a locally optimal tour has length at most avg_I for the 2-change neighborhood function, but that it may have length larger than avg_I for the node-insertion neighborhood function.

a) Prove that

$$avg_I = \frac{2}{n-1} \sum_{\substack{i,j \in C \\ i \neq j}} d_{ij},$$

where $C = \{0, 1, \ldots, n-1\}$ denotes the set of cities in problem instance I.

For a given tour τ and a pair $e_i = \{\tau(i), \tau(i+1)\}$ and $e_j = \{\tau(j), \tau(j+1)\}$ of non-adjacent edges in τ, we define $\Delta_\tau(e_1, e_2)$ as the change in tour length resulting from a 2-change that replaces e_i and e_j with two new edges. It can be verified that

$$\Delta_\tau(e_1, e_2) = d_{\tau(i),\tau(j)} + d_{\tau(i+1),\tau(j+1)} - d_{\tau(i),\tau(i+1)} - d_{\tau(j),\tau(j+1)}.$$

For a given edge e_i we define $V_\tau(e_i)$ as the set of edges in τ that are not adjacent to e_i. Furthermore, we define

$$\Delta_\tau(e_i) = \sum_{e_j \in V_\tau} \Delta_\tau(e_i, e_j)$$

and

$$\delta_k = \sum_{\substack{j \in C \\ j \neq k}} d_{kj}.$$

b) Prove that each edge $e_i = \{\tau(i), \tau(i+1)\}$ in τ satisfies

$$\Delta_\tau(e_i) = \delta_{\tau(i)} + \delta_{\tau(i+1)} - (n-2)d_{\tau(i),\tau(i+1)} - f(\tau).$$

c) Prove that

$$\sum_{e \in \tau} \Delta_\tau(e) = 4 \sum_{\substack{i,j \in C \\ i \neq j}} d_{ij} - (2n-2)f(\tau).$$

Let $\hat{\tau}$ be a locally optimal tour with respect to the 2-change neighborhood function.

d) Use (c) to prove that $f(\hat{\tau}) \leq avg_I$.

In the remainder of this exercise we will prove that the length of a locally optimal tour with respect to the node-insertion neighborhood function may be larger than avg_I. Consider the problem instance depicted in Figure 5.12. The problem instance consists of ten cities numbered from 0 to 9 and it has a distance matrix

Figure 5.12. Problem instance discussed in Exercise 5. Edges with length 0 have been omitted.

that is defined by

$$d_{ij} = \begin{cases} 1 & \text{if } j = (i+1) \bmod 10, \\ 3 & \text{if } j = (i+2) \bmod 10, \\ 0 & \text{otherwise.} \end{cases}$$

e) Prove that tour $\hat{\tau} = (0, 1, \ldots, 9)$ is locally optimal with respect to the node-insertion neighborhood function.

f) Prove that tour $\hat{\tau}$ defined in (e) satisfies $f(\hat{\tau}) > avg_I$.

6. Let I be an arbitrary problem instance of EUCLIDEAN TSP, and let I' be the problem instance obtained from I by replacing the distance matrix d by d' with $d'_{ij} = d^2_{ij}$.

a) Prove or disprove the following claim: if τ^* is an optimal tour for I, then it is also an optimal tour for I'.

The following exercises are concerned with SYMMETRIC TSP.

b) Indicate how to adapt the proof of Corollary 5.1 to turn it into a proof of the result that, unless P = NP, no polynomially searchable neighborhood exists for SYMMETRIC TSP with a performance bound of $2^{p(n)}$ for some fixed polynomial p.

c) Give a variable-depth algorithm for SYMMETRIC TSP that is based on the 2-change neighborhood function.

d) What can you say about the performance ratio of the neighborhood function of (c)?

7. In Exercise 2 we defined MAXIMUM CLAUSE-WEIGHTED SATISFIABILITY. Let MAXIMUM SATISFIABILITY be the unweighted variant of this problem, i.e., in MAXIMUM SATISFIABILITY (MAX-SAT)[1] we are asked to maximize the number of satisfied clauses. Consider the following two decision problems.

Definition 5.11 [SATISFIABILITY]. Given are a set U of binary variables and a set C of clauses over U. A clause is satisfied by a given truth assignment if and

[1] In the literature MAX-SAT can also refer to the weighted variant of the problem.

only if at least one of its literals is true. Does there exist a truth assignment that satisfies all clauses? □

Definition 5.12 [MAX-SAT SUBOPTIMALITY]. Given are an instance I of MAX-SAT that consists of m clauses and a truth assignment t for I that satisfies $m - 1$ clauses. Is t suboptimal, i.e., does there exist a truth assignment that satisfies all m clauses? □

a) SATISFIABILITY is known to be NP-complete. Prove that MAX-SAT SUB-OPTIMALITY is NP-complete by giving a reduction from this NP-complete problem.

b) What is the relation of this result to local search?

8. Consider algorithm $\mathcal{A}_\varepsilon(N)$ described in Figure 5.11, and suppose that we use an R'-approximation algorithm to derive an initial solution for $\mathcal{A}_\varepsilon(N)$, where $R' > R$. How does this affect the time complexity of the algorithm?

6

Time Complexity

One way of tackling an intractable NP-hard problem is by applying local search, in particular, iterative improvement. Instead of solving the problem to optimality, iterative improvement only provides a local optimum. In other words, the algorithm solves the relaxed problem in which the goal is not to find a global optimum, but to find a local optimum. This local search problem is formalized by the following two definitions.

Definition 6.1. An *instance* of a local search problem is a triple (S, f, N), where (S, f) is an instance of a combinatorial optimization problem and $N : S \to 2^S$ is a neighborhood function on S. □

Definition 6.2. A *local search problem* Π_{LS} is specified by a set of problem instances and it is either a minimization or a maximization problem. The problem is to find for a given instance (S, f, N) a locally optimal solution $\hat{s} \in S$. For a minimization problem, this means that $f(\hat{s}) \le f(s)$ for all $s \in N(s)$, and for a maximization problem it means that $f(\hat{s}) \ge f(s)$ for all $s \in N(s)$. □

To guarantee that iterative improvement is suitable for tackling problem instances of realistic size, it is important to determine its time complexity. If the running time is bad, i.e., exponential, then an immediate question is whether this large running time is caused by the simplicity of iterative improvement or by the inherent intractability of the local search problem it solves. Note that a local search problem cannot be harder than the underlying combinatorial optimization problem. In this chapter we both study the time complexity of iterative improvement and the computational

complexity of local search problems. We assume the reader to have some experience in applying the theory of NP-completeness. For a brief overview we refer to Appendix B.

We start in Section 6.1 by studying the computational complexity of local search problems. The section is based on the paper of Yannakakis [1997], who gives an excellent presentation of the complexity theory for local search problems introduced by Johnson, Papadimitriou & Yannakakis [1988]. In Section 6.2 we then discuss the time complexity of iterative improvement for several local search problems. We stress that if iterative improvement has a polynomial running time for a local search problem, then this local search problem is easy by definition. However, if iterative improvement has an exponential running time, then this need not mean that it solves a hard local search problem.

6.1 Computational Complexity of Local Search Problems

To distinguish easy *decision* problems from hard ones, the theory of NP-completeness has been developed. In a nutshell, the idea of this theory is as follows. A class NP of decision problems is defined that contains the decision variants of many apparently hard and interesting combinatorial optimization problems. To relate the complexity of two problems in NP, polynomial-time reductions are introduced. If a problem Π is polynomially reducible to problem Π', then Π' is at least as hard as Π, which means that a polynomial-time algorithm for Π' can only exist if it exists for Π. The problems in NP to which all problems in this class are polynomially reducible are called NP-complete. These problems are the hardest problems in NP since they only allow an efficient polynomial-time algorithm if such an algorithm exists for all problems in NP. This is very unlikely as no polynomial-time algorithm has been found for an NP-complete problem despite the enormous amount of effort that has been spent by many experts on finding one.

A similar strategy is used to identify hard local search problems for which it is very unlikely that they can be solved by a polynomial-time algorithm. First of all, we define a complexity class, called PLS, which contains local search problems that are specified by a combinatorial optimization problem from NPO and a 'reasonable' corresponding neighborhood function; see Appendix B for the definition of NPO. Next, we present a reduction to express that a local search problem is at least as hard as another given local search problem. Using this reduction, we can identify the hardest problems in PLS. Because no polynomial-time algorithm is known for any of these PLS-complete problems, as they are called, it is commonly believed that no such algorithm exists for these problems.

We now further elaborate on this theory. Let Π be a combinatorial optimization problem from NPO and let N be a neighborhood function for it. A successful application of local search for this problem depends on whether it is easy to generate an initial solution and to check whether a solution is locally optimal and, if not, to derive a better neighbor. In terms of iterative improvement, this means that both the initialization and each single iteration of the algorithm can be implemented to run in polynomial time. The number of iterations, however, may be exponential.

The class PLS contains all local search problems in which the properties described are satisfied. Note that we have already encountered the properties in Sections 5.3 and 5.4.

Definition 6.3. Let Π_{LS} be a local search problem and let Π be the underlying combinatorial optimization problem. Local search problem Π_{LS} is in the class *PLS (Polynomial-time Local Search)* if $\Pi \in$ NPO and if two polynomial-time algorithms A and B exist that satisfy the following properties.

- For a problem instance (S, f, N) of Π_{LS}, algorithm A returns a solution $s \in S$.
- For a problem instance (S, f, N) of Π_{LS} and a solution $s \in S$, algorithm B decides whether s is a local optimum and, if this is not the case, it returns a neighboring solution with better cost.

\square

The following definition specifies a PLS-reduction by which we can relate the complexity of two local search problems. From the definition it follows directly that if local search problem Π_{LS} is PLS-reducible to local search problem Π'_{LS}, then the existence of a polynomial-time algorithm for Π'_{LS} implies the existence of such an algorithm for Π_{LS}. In other words, Π'_{LS} is at least as hard as Π_{LS}. Furthermore, it can be verified that PLS-reductions are transitive, which means that if Π_{LS} is PLS-reducible to Π'_{LS} and Π'_{LS} is PLS-reducible to Π''_{LS}, then Π_{LS} is also PLS-reducible to Π''_{LS}.

Definition 6.4. Local search problem Π_{LS} is *PLS-reducible* to local search problem Π'_{LS}, denoted by $\Pi_{LS} \propto \Pi'_{LS}$, if two polynomial-time algorithms φ_1 and φ_2 exist that satisfy the following properties; see also Figure 6.1.

- Algorithm φ_1 transforms a problem instance I of Π_{LS} into a problem instance $\varphi_1(I)$ of Π'_{LS}.
- Algorithm φ_2 maps a problem instance $I = (S, f, N)$ of Π_{LS} and a solution $s' \in S'$ with $\varphi_1(I) = (S', f', N')$ to a solution $s \in S$.
- For a problem instance I of Π_{LS}, we have that if $s' \in S'$ is a local optimum for $\varphi_1(I) = (S', f', N')$, then $\varphi_2(I, s')$ is a local optimum for I.

The pair (φ_1, φ_2) is called a *PLS-reduction*. \square

We note that the definition of φ_2 can be weakened. For φ_2 it suffices that instead of all solutions in S', only the local optima are mapped to a solution in S. However, if we have $\Pi_{LS} \in$ PLS, then we can easily transform such a φ_2 into an algorithm that satisfies Definition 6.4 by mapping all solutions that are not in the domain of the restricted φ_2 to the solution returned by the polynomial-time algorithm A defined in Definition 6.3. Analogously to NP-completeness, we define the notion of PLS-completeness to indicate the hardest problems in PLS.

Definition 6.5. A local search problem in PLS is *PLS-complete* if each problem $\Pi_{LS} \in$ PLS is PLS-reducible to it. \square

$$S \qquad\qquad S'$$

Figure 6.1. Example of algorithm φ_2 for mapping solutions from S' to solutions from S. The squares indicate the local optima.

In Sections 6.1.1 we prove our first PLS-complete problem, called CIRCUIT/flip. Once we have this result we only have to prove for a given problem Π_{LS} that CIRCUIT/flip is PLS-reducible to it to conclude that Π_{LS} is PLS-complete. This is used in Section 6.1.2 to prove that UNIFORM GRAPH PARTITIONING with the Kernighan-Lin neighborhood function presented in Section 2.5 is PLS-complete. We note that PLS-reductions used for proving PLS-completeness results are generally much more intricate than the polynomial-time reductions used for proving NP-completeness results. To gain an impression of how a PLS-reduction works, we recommend the reader to study at least the proof of Theorem 6.5, in which we present a relatively simple and elegant PLS-reduction. In Appendix C we give a list of local search problems that have been proven to be PLS-complete.

Let NPCO contain all NP-hard problems from NPO. When determining the complexity of solving some combinatorial optimization problem from NPO to optimality, one generally aims to decide whether the problem is in PO or in NPCO. It is therefore interesting to relate the class PLS and the subclass of PLS-complete problems to the classes PO, NPO, and NPCO. This is the focus of the remainder of this section.

First we show that PLS \subseteq NPO. Let Π_{LS} be an arbitrary local search problem in PLS. Problem Π_{LS} can be formulated as a minimization problem $\Pi \in$ NPO by defining an instance $(S, f') \in \Pi$ for any instance $(S, f, N) \in \Pi_{LS}$, such that cost function $f' : S \to \{0,1\}$ satisfies $f'(s) = 0$ if s is locally optimal and $f'(s) = 1$ otherwise. The existence of algorithm B as defined in Definition 6.3 for Π_{LS} implies that f' is computable in polynomial time. From this it follows that Π is indeed a problem from NPO. Moreover, Π_{LS} and Π are clearly equivalent. As a result, we get PLS \subseteq NPO.

We now show that PO \subseteq PLS. Let $\Pi \in$ PO and let \mathcal{A} be an optimal polynomial-time algorithm for Π. Without loss of generality we assume that Π is a minimization problem. Then we define minimization problem $\Pi' \in$ PO such that any instance $I = (S, f)$ of Π corresponds to an instance (S, f') of Π', where $f'(s) = 0$ if solution s

is optimal with respect to cost function f and $f'(s) = 1$ otherwise. If we define a neighborhood function for Π' in such a way that each solution has only one neighbor, namely the solution returned by \mathcal{A}, then we obtain a local search problem that is equivalent to Π. Furthermore, the problem is in PLS. Algorithm A in Definition 6.3 can be defined as \mathcal{A} and algorithm B in this definition can be realized by first comparing the cost of the solution given by \mathcal{A} with the cost of a given solution s and by then returning the optimal solution derived by \mathcal{A} if the latter cost is larger. Hence, we now obtain the following result.

Theorem 6.1. *The complexity classes PO, PLS, and NPO satisfy*

$$PO \subseteq PLS \subseteq NPO.$$

\square

From the relation PLS \subseteq NPO it follows that a PLS-complete problem is not harder than any NP-hard problem in NPO, as will now be shown. Suppose that a polynomial-time algorithm exists for an NP-hard problem in NPCO. We then have P = NP. By Theorem B.2 in Appendix B this yields that all problems in NPO are polynomially solvable, including those in PLS. To substantiate the intractability of PLS-complete problems, a more interesting question is whether PLS-complete problems are also at least as hard as the problems in NPCO. If the answer were to be affirmative, then the existence of a polynomial-time algorithm for a PLS-complete problem would be as unlikely as the existence of such an algorithm for a problem in NPCO. However, this is not the case because the following theorem states that if NP \neq co-NP, then a PLS-complete problem is not NP-hard. Hence, in the unlikely case that PLS-complete problems turn out to be polynomially solvable, this still does not imply that the problems in NPCO are also solvable in polynomial time. Figure 6.2 depicts the relations between the different complexity classes.

Theorem 6.2. *If NP \neq co-NP, then none of the problems in PLS are NP-hard.*

Proof. The result can be proved in a similar way as Theorem 5.9. Note that Theorem 5.9 easily follows from this theorem. Suppose that a problem $\Pi_{LS} \in$ PLS is NP-hard. We show that this contradicts NP \neq co-NP by showing that it implies that, for any decision problem Π_D in NP, the complementary problem Π_D^c obtained by reversing the answers in Π_D is also in NP.

As Π_{LS} is NP-hard, a polynomial-time algorithm \mathcal{A} exists that decides Π_D, where \mathcal{A} may use an oracle that, for any problem instance of Π_{LS}, returns a local optimum. Hence, if for a problem instance I of Π_D we are given a sequence consisting of local optima that may be returned successively by the oracle during the execution of \mathcal{A} on I, then we can determine in polynomial time whether I is a yes-instance or a no-instance of Π_D. As a result, we can use this sequence of locally optimal solutions as a certificate for yes-instances of Π_D^c. Furthermore, the certificate, which is of polynomial size, can be checked in polynomial time for validity by substituting each call of the oracle in \mathcal{A} by a polynomial-time procedure that checks whether the corresponding solution in the certificate is a solution to the problem instance of Π_{LS} given to the oracle. The existence of polynomial-time algorithm

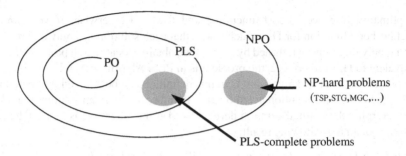

Figure 6.2. Positioning of the classes PO, PLS, and NPO in the case that P \neq NP, NP \neq co-NP, and PO \neq PLS.

B in Definition 6.3 shows that this polynomial-time procedure exists. This proves $\Pi_D^c \in$ NP, which completes the proof of the theorem. \square

6.1.1 *First PLS-Complete Problem: A Starting Point

Suppose we want to apply local search to the problem of finding an integer s with $0 \leq s < 2^n$ for which a given function $f : \{0, 1, \ldots, 2^n - 1\} \to \{0, 1, \ldots, 2^m - 1\}$ is minimized. We can use the neighborhood function in which $s' \in \{0, 1, \ldots, 2^n - 1\}$ is a neighbor of s if s' can be obtained from s by flipping exactly one of the n bits in the binary representation of s. The corresponding local search problem, formulated in terms of Boolean circuits, is the first problem that has been proved to be PLS-complete.

Boolean circuits are theoretical counterparts of the digital circuits from which computers are made. They compute Boolean functions $f : \{0, 1\}^n \to \{0, 1\}^m$ and, conversely, each Boolean function is computed by a circuit.

Definition 6.6. A *Boolean circuit* is a directed acyclic graph $D = (V, A)$. The node set V consists of n input nodes and $|V| - n$ gates. The input nodes have indegree zero and are labeled by the binary variables x_1, x_2, \ldots, x_n. The labels of the gates are taken from the set $\{\wedge, \vee, \neg\}$ of Boolean functions. The gates with outdegree zero are called the output nodes and they are additionally labeled by the binary variables y_1, y_2, \ldots, y_m. Boolean circuit D computes a Boolean function $f : \{0, 1\}^n \to \{0, 1\}^m$ by deriving for given values of the input variables x_1, x_2, \ldots, x_n corresponding values for the output variables y_1, y_2, \ldots, y_m in the following way.

Let l be the label of a gate g. If l is given by \neg, then the value of g is one minus the value of the node from which the only incoming edge of g is incident. Next, suppose that $l \in \{\wedge, \vee\}$. Then g has exactly two incoming edges. Let o_1, o_2 be the values of the nodes from which these two edges are incident. If g is labeled \wedge, then it is assigned the value one if $o_1 = o_2 = 1$ and it is assigned the value zero, otherwise. If g is labeled \vee, then it is assigned the value zero if $o_1 = o_2 = 0$ and it is assigned the value one otherwise. The *size* of Boolean circuit $D = (V, A)$ is given by the number of nodes in V. \square

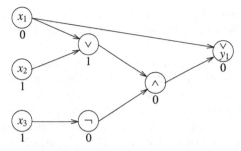

Figure 6.3. Boolean circuit computing the function $((x_1 \lor x_2) \land \neg x_3) \lor x_1$. The numbers below the nodes indicate their value in the case where the input is given by $(0, 1, 1)$.

Example 6.1. Figure 6.3 depicts a Boolean circuit. It consists of three input nodes (x_1, x_2, x_3) and four gates. Of the four gates, only the rightmost node is an output node. In the figure we indicate the values of the nodes in the circuit for the case where the input is given by $(x_1, x_2, x_3) = (0, 1, 1)$. Obviously, the circuit computes the Boolean function $((x_1 \lor x_2) \land \neg x_3) \lor x_1$. □

By interpreting the input and output vectors as two numbers written in binary notation, i.e., by interpreting the input as $\sum_{i=1}^{n} 2^{i-1} x_i$ and the output as $\sum_{i=1}^{m} 2^{i-1} y_i$, a Boolean circuit can compute an integer function. Hence, the local search problem mentioned above can be formalized by the following definition. We note that the Hamming distance of two strings s and s' of equal size, denoted by $H(s, s')$, is given by the number of positions in which s and s' differ.

Definition 6.7 [CIRCUIT/flip]. Given is a Boolean circuit D with n input nodes x_1, x_2, \ldots, x_n and m output nodes y_1, y_2, \ldots, y_m. The neighborhood function, called flip, is based on the solution space $\{0, 1\}^n$ containing all possible input vectors (x_1, x_2, \ldots, x_n) of the circuit. A solution s is a neighbor of solution s' if s and s' have a Hamming distance $H(s, s')$ of one. Find a solution \hat{s} that is locally optimal with respect to the cost function

$$f(\hat{s}) = \sum_{i=1}^{m} 2^{i-1} y_i,$$

which gives the integer interpretation of the output of Boolean circuit D on input \hat{s}. In MIN-CIRCUIT/flip it is our goal to minimize the cost function, while in MAX-CIRCUIT/flip it is our goal to maximize the cost function. □

It can easily be verified that both MIN-CIRCUIT/flip and MAX-CIRCUIT/flip belong to the class PLS. The following lemma states that they are also equally hard.

Lemma 6.1. MIN-CIRCUIT/*flip and* MAX-CIRCUIT/*flip are PLS-reducible to each other.*

Proof. We only define a PLS-reduction (φ_1, φ_2) from MIN-CIRCUIT/flip to MAX-CIRCUIT/flip. A PLS-reduction from MAX-CIRCUIT/flip to MIN-CIRCUIT/flip can be defined similarly.

Let Boolean circuit D define an arbitrary problem instance of MIN-CIRCUIT/flip, and let f be the cost function computed by D. Algorithm φ_1 constructs an instance of MAX-CIRCUIT/flip, i.e., a Boolean circuit $\varphi_1(D)$, as follows. For each output node v in D algorithm φ_1 adds a new node v' that is connected to D via an arc from v to v' and that has label \neg. This gives Boolean circuit $\varphi_1(D)$ that flips the value of each output node of D. Hence, $\varphi_1(D)$ computes $2^m - 1 - f(s)$ on input $s \in \{0, 1\}^n$. Obviously, the local and global optima of D for MIN-CIRCUIT/flip are the same as the local and global optima of $\varphi_1(D)$ for MAX-CIRCUIT/flip. Hence, (φ_1, φ_2) defines a valid PLS-reduction, where algorithm φ_1 is as defined above and φ_2 is the identity, i.e., φ_2 returns s for all $s \in \{0, 1\}^n$. □

The following theorem is useful for proving that MIN-CIRCUIT/flip and MAX-CIRCUIT/flip are PLS-complete. We note that $\{0, 1\}^+$ represents the set of all binary vectors (or strings) containing one or more entries.

Theorem 6.3. *If $f : \{0, 1\}^+ \to \{0, 1\}^+$ is a function that is computable in polynomial time, then a polynomial-time algorithm exists, which for any n derives a Boolean circuit that computes $f(s)$ for any string $s \in \{0, 1\}^n$ of length n. As the algorithm runs in polynomial time, the derived Boolean circuits are of polynomial size.*

Proof. Suppose that for some given language L we have an algorithm that, when implemented on a Turing machine, decides L in $\mathcal{O}(t(n))$ time, where n denotes the length of the input string. Then, Theorem 9.25 of Sipser [1997] states that for each n a Boolean circuit exists of size $\mathcal{O}(t^2(n))$ that decides L on inputs of length n. He proves this result by showing how a Boolean circuit can be constructed that simulates the computation of the Turing machine on inputs of a fixed length. This construction can be done in time polynomial in $t(n)$. A similar proof can be used to show that this theorem holds. □

We are now able to prove our first PLS-complete problem. Although the result is rather impressive, the proof is relatively simple.

Theorem 6.4. MIN-CIRCUIT/*flip and* MAX-CIRCUIT/*flip are both PLS-complete.*
Proof. Let Π_{LS} be an arbitrary minimization problem in PLS. We show that Π_{LS} is PLS-reducible to MIN-CIRCUIT/flip. In a similar way it can be proved that any maximization problem in PLS is PLS-reducible to MAX-CIRCUIT/flip. As by Lemma 6.1 MIN-CIRCUIT/flip and MAX-CIRCUIT/flip are PLS-reducible to each other, this implies that both problems are PLS-complete.

Preliminaries. We first introduce some notation and some assumptions we can make regarding local search problem Π_{LS}. We can assume that problem instances and solutions are described by strings over the alphabet $\{0, 1\}$. The encoding of a problem instance I as a string in $\{0, 1\}^+$ is denoted by e_I. Remember that the size $|I|$ of

instance I is given by the number of bits in e_I. Analogously, e_s denotes the encoding of a solution s.

The underlying combinatorial optimization problem of a problem in PLS is by definition in NPO. Furthermore, the size of a solution of a problem in NPO is, also by definition, polynomially bounded in the input size. This means that a polynomial p exists, such that the size of a solution of an arbitrary problem instance I of Π_{LS} is bounded from above by $p(|I|)$. Hence, since we are allowed to switch to a different encoding scheme for solutions as long as the resulting increase in the size of the solutions is polynomially bounded in $|I|$, we may assume without loss of generality that

- a polynomial p exists, such that the size of a solution of an arbitrary problem instance I is exactly $p(|I|)$ and

- for any two different solutions s and s' of some problem instance I, the Hamming distance between e_s and $e_{s'}$ is at least two.

The first property can be ensured by adding a one behind the encoding of a solution s followed by $p(|I|) - e_s - 1$ zeros, where $p(|I|)$ is strictly larger than the size of any solution of I, i.e., at least the one is appended behind a solution. The original encoding can easily be derived from the new one by removing the uniquely defined tail $100\ldots0$. The second property can be ensured by duplicating each bit in the encoding of a solution. In the following, we write p instead of $p(|I|)$ if I follows from the context.

Proof strategy. We now return to proving $\Pi_{LS} \propto$ MIN-CIRCUIT/flip. Instead of deriving this result directly, we prove it via an intermediate artificial problem. This means that we define a local search problem Π and prove $\Pi_{LS} \propto \Pi \propto$ MIN-CIRCUIT/flip. Problem Π contains the same set of problem instances as Π_{LS}, and it is formulated in terms of the encoding of its problem instances. For a problem instance I with encoding e_I it asks for a fixed-length binary string $s \in \{0,1\}^{2p+2}$ that minimizes $f(e_I, s)$, where $f : \{0,1\}^+ \to \mathbb{N}$ or, equivalently, $f : \{0,1\}^+ \to \{0,1\}^+$, is a polynomially computable cost function that will be defined below. Furthermore, Π has the same solution space and neighborhood function as MIN-CIRCUIT/flip. This means that the solution space is given by all strings from $\{0,1\}^{2p+2}$ and that solutions s and s' are neighbors of each other if and only if $H(s, s') = 1$. Obviously, we have $\Pi \in$ PLS. We first prove $\Pi \propto$ MIN-CIRCUIT/flip. Afterwards, we show that also $\Pi_{LS} \propto \Pi$ holds, which yields $\Pi_{LS} \propto$ MIN-CIRCUIT/flip.

Proof of $\Pi \propto$ MIN-CIRCUIT/flip. Consider local search problem Π. As f is polynomially computable, Theorem 6.3 yields that a polynomial-time algorithm \mathcal{A} exists that for any n derives a polynomial-sized Boolean circuit that computes f for strings of length n. We now define algorithm φ_1 in the PLS-reduction (φ_1, φ_2) from Π to MIN-CIRCUIT/flip as follows. For any problem instance I of Π, it first runs \mathcal{A} with input $n = |I| + 2p + 2$. This gives a Boolean circuit for computing f for any problem instance with the same size as I. By substituting the bits of e_I in the first $|I|$ inputs of the circuit, we obtain a Boolean circuit that computes $f(e_I, s)$ for any solution $s \in \{0,1\}^{2p+2}$. This Boolean circuit is returned by φ_1. We note that an input node

of a Boolean circuit is by definition a variable and not a constant. However, it is easy to eliminate input nodes with label 0 or 1 from the circuit. In this way, φ_1 defines a problem instance of MIN-CIRCUIT/flip with the same global and local optima as problem instance I of Π. In other words, (φ_1, φ_2) defines a PLS-reduction if φ_1 is as defined above and φ_2 is given by the identity.

Proof of $\Pi_{LS} \propto \Pi$. In the remainder of the proof we construct a PLS-reduction (φ_1, φ_2) from Π_{LS} to Π, where we still have the freedom to fill in the details in the definition of the cost function f of Π. We only have to take care that f is computable in polynomial time. As mentioned above, Π_{LS} and Π have the same set of problem instances. Therefore, it is not surprising that we simply define φ_1 as the identity. Nevertheless, we still write both I and $\varphi_1(I)$ to distinguish between whether I refers to a problem instance of Π_{LS} or to a problem instance of Π. We now discuss algorithm φ_2. A solution s of $\varphi_1(I)$ is a binary string of length $2p+2$ and is considered to consist of three parts. The first two parts are both of length p and refer to encodings of solutions of Π_{LS}. The last part consists of only two bits and has the interpretation of the state of solution s. We define $R = \{uu00 \mid u$ is the encoding of a solution of $I\}$ as a subset of the solutions of $\varphi_1(I)$. Based on the obvious one-to-one correspondence between R and the solutions of I, we let φ_2 return the solution with encoding u on input $uu00 \in R$ and solution $A(I)$ otherwise, where algorithm A is as defined in Definition 6.3. This concludes the definition of the PLS-reduction (φ_1, φ_2). In the remainder of this proof, we mean the encoding e_u of solution u when we refer to a solution u of I. By using this convention, we can for instance write $R = \{uu00 \mid u$ is a solution of $I\}$.

We now show that the described PLS-reduction is valid, where we exploit the freedom to choose the cost function f. Let $T_{\varphi_1(I)}$ be the transition graph of $\varphi_1(I)$. We say that a path from $s \in R$ to $s' \in R$ in $T_{\varphi_1(I)}$ is direct if no intermediate solutions in this path are in R. We define f in such a way that the transition graph satisfies the following two conditions.

- *Condition 1.* $T_{\varphi_1(I)}$ contains a direct path from $s = uu00 \in R$ to $s' = ww00 \in R$ if and only if $w = B(u)$, where B is as defined in Definition 6.3; see Figure 6.4. The 'if' part implies the following. Let γ be a sequence of solutions for I that are successively obtained in an execution of the iterative improvement algorithm if it uses algorithm B in each iteration to move to a better neighbor. Then, a path in $T_{\varphi_1(I)}$ exists for which it holds that the removal of all solutions that are not in R results in the sequence of solutions from R that corresponds to sequence γ.

- *Condition 2.* For any solution $s \notin R$, transition graph $T_{\varphi_1(I)}$ contains a path to a solution $s' \in R$.

The first condition yields that if $s = uu00 \in R$ is a local optimum for $\varphi_1(I)$, then $\varphi_2(I,s) = u$ is a local optimum for I. Furthermore, it follows from the second condition that the local optima of $\varphi_1(I)$ form a subset of R. Hence, if s is locally optimal for $\varphi_1(I)$, then $\varphi_2(I,s) = u$ is locally optimal for I. As a result, proving the two con-

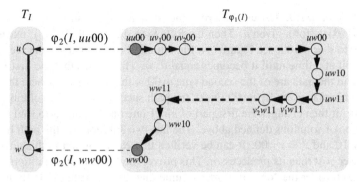

Figure 6.4. The correspondence between an arc (u,w) in transition graph T_I with $w = B(u)$ and a direct path in transition graph $T_{\varphi_1(I)}$ from the solutions $uu00 \in R$ to $ww00 \in R$. The only depicted solutions from R are the dark shaded ones, i.e., $uu00$ and $ww00$.

ditions suffices to show that (φ_1, φ_2) is a valid PLS-reduction. We note that for our purpose we do not need to prove the only-if part of the first condition. However, this result will be useful later on when we introduce tight PLS-reductions to analyze the time complexity of iterative improvement.

Definition of cost function f. It now remains to be shown that a cost function f can be constructed, such that the above two conditions are satisfied. To define f, we need the following definition. Let u and w with $u \neq w$ be two arbitrary binary strings of length p. Then we can change u into w by flipping $H(u,w)$ bits in u. We define $V(u,w)$ as the set of $H(u,w)$ strings that are obtained if we change u into w by flipping the bits in which u differs from w in increasing order of their position. By definition $V(u,w)$ contains w but not u. Below we define $f(s)$ for some of the strings $s \in \{0,1\}^{2p+2}$. These strings are called well-structured. All other strings are called unstructured. In the definition of f, we let $u \in \{0,1\}^p$ represent a solution of I and we let $g(I,u)$ be defined as the cost of solution u in problem instance I of Π_{LS}.

- $f(e_I, uu00) = (2p+4)g(I,u)$,
- $f(e_I, uv00) = (2p+4)g(I,w) + (p+2) + H(v,w) + 2$, where $v \in V(u,w)$ with $w = B(u)$,
- $f(e_I, uw10) = (2p+4)g(I,w) + (p+2) + 1$ with $w = B(u)$,
- $f(e_I, vu11) = (2p+4)g(I,u) + H(u,v) + 2$, where $v \in \{0,1\}^p$ is an arbitrary string of length p, and
- $f(e_I, uu10) = (2p+4)g(I,u) + 1$.

For any other (unstructured) string s, we define the cost $f(e_I, s)$ such that it is larger than the cost of any well-structured string. Note that this implies that transition graph $T_{\varphi_1(I)}$ does not contain an arc from a well-structured solution to an unstructured solution. Details are provided after the proof of Condition 1.

Proof of Condition 1. Let us first prove the 'if' part of Condition 1. Suppose that $w = B(u)$ exists for a given u. Then the following direct path in $T_{\varphi_1(I)}$ exists from $s = uu00$ to $s' = ww00$; see Figure 6.4. First of all, the path changes the second part of s one bit at a time until it becomes equal to w. This means that the first $H(u,w)$ solutions in the path are of the second type $uv00$ with $v \in V(u,w)$, where the last of these solutions is given by $uw00$. Next, the path successively visits solutions $uw10$ and $uw11$. It then changes the first part of $uw11$ one bit at a time into $ww11$ via the fourth type of solutions defined above. The last two solutions on the path from s to s' are $ww10$ and $s' = ww00$. It can be verified that each solution in this direct path has a lower cost than its predecessor. This proves the 'if' part of Condition 1.

Next, consider the 'only-if' part. Assume that a direct path exists from s to s'. Then this path must have the form that we just described, i.e., it first changes the second part of $s = uu00$ into w with $w = B(u)$ and next it transforms $uw00$ via $uw11$ into $s' = ww00$. This can be verified as follows. Changing the state of $s = uu00$ cannot improve the cost of a solution. Furthermore, changing one bit in the first part u of s transforms it into a string v that is not a solution of I because by assumption any two solutions in I have Hamming distance of at least two. As a result, we can only obtain a solution with a lower cost than s if we change its second part. This can only be done via the second type of solution $uv00$ with $v \in V(u,w)$, where $w = B(u)$. Using similar arguments it can be shown that the remainder of the path also has to be of the described format. This proves that Condition 1 indeed holds.

Proof of Condition 2. Consider the second condition, which states that for any $s \notin R$, transition graph $T_{\varphi_1(I)}$ contains a path from s to a solution $s' \in R$. The condition is satisfied for well-structured solutions as they are all on a path in $T_{\varphi_1(I)}$ from a string $s \in R$ to some other string $s' \in R$. We now show that it is also satisfied for any unstructured solution s. Regarding the cost of an unstructured solution, we only assumed that it is larger than the cost of any well-structured solution. As the cost $g(I,u)$ of a solution u in I is computable in time that is polynomial in $|I|$, a polynomial q exists such that for any problem instance I and solution u of I the cost $g(I,u)$ can be encoded in $q(|I|)$ bits. Hence, we have $g(I,u) \leq 2^{q(|I|)}$, which implies that the cost of any well-structured solution of $\varphi_1(I)$ is bounded from above by $Z = (4p + 1)2^{q(|I|)} + p + 2$. For an unstructured solution s we now define $f(e_I, s)$ as $Z + H(s, aa00)$, where $a = A(I)$ is the solution returned by algorithm A of Π_{LS}. Then the cost of unstructured solution s is indeed larger than the cost of any well-structured solution, and a path in $T_{\varphi_1(I)}$ exists from s to $aa00 \in R$.

It can be verified that, as g is computable in polynomial time and A and B are polynomial-time algorithms, cost function f is computable in polynomial time. This concludes the proof of the theorem. $\qquad\square$

6.1.2 *UGP with Kernighan-Lin Neighborhood Function is PLS-Complete

Obviously, it is much easier to show for a given local search problem $\Pi_{LS} \in$ PLS that a specific problem in PLS is PLS-reducible to it than to show that all problems in PLS are PLS-reducible to it. Therefore, we took an important step in Section 6.1.1 by proving that CIRCUIT/flip is PLS-complete. It now suffices to prove

CIRCUIT/flip $\propto \Pi_{LS}$ to conclude that Π_{LS} is PLS-complete. Note that this strategy is also used to prove that a decision problem $P \in$ NP is NP-complete: one generally tries to define a polynomial-time reduction from a known NP-complete decision problem Q to P.

In Section 2.5 we discuss the Kernighan-Lin neighborhood function as an effective generalization of the swap neighborhood function for UNIFORM GRAPH PARTITIONING (UGP). The main result of this section is that the local search problem of determining a local optimum with respect to the Kernighan-Lin neighborhood function is PLS-complete. By a much more involved proof it has been shown that the simpler swap neighborhood function also implies a PLS-complete problem. Note that this is a stronger result because a locally optimal solution with respect to the Kernighan-Lin neighborhood function is also swap optimal.

We first introduce a problem, called POSITIVE NOT-ALL-EQUAL MAX-3SAT (POS NAE MAX-3SAT), together with a neighborhood function for it and prove that the corresponding local search problem is PLS-complete. Using this auxiliary result we prove our claimed main result. In itself, the problem POS NAE MAX-3SAT does not have much practical meaning. Let us discuss briefly the terminology used in the problem definition of POS NAE MAX-3SAT. Let U be a set of binary variables. A truth assignment for U assigns to each variable in U either the value 0 (false) or 1 (true). A variable $u \in U$ is called a positive literal, and its negation $\bar{u} = 1 - u$ is called a negative literal. A clause is a set of literals and/or constants, where a constant has the value either 0 or 1.

Definition 6.8 [POSITIVE NOT-ALL-EQUAL MAX-3SAT (POS NAE MAX-3SAT)]. Given is a set U of binary variables, a set C of clauses over U, and a positive integer weight $w(c)$ for each clause $c \in C$. A clause does not contain negative literals, and the number of positive literals plus the number of constants in each clause is at most three. A clause is satisfied by some truth assignment if and only if it contains at least one literal/constant with value one and at least one literal/constant with value zero. The problem is to find a truth assignment that maximizes the sum of the weights of clauses that are satisfied. □

An obvious neighborhood function for POS NAE MAX-3SAT is the flip neighborhood function in which truth assignment t' is a neighbor of truth assignment t if t' can be obtained from t by changing (flipping) the value of one variable. In much the same way as the Kernighan-Lin neighborhood function for UGP is obtained by applying variable-depth search to the swap neighborhood function, the Kernighan-Lin neighborhood function for POS NAE MAX-3SAT is obtained by applying variable-depth search to the flip neighborhood function. This means that a truth assignment t' is a neighbor of truth assignment t with respect to the Kernighan-Lin neighborhood function if it can be derived from t by performing a sequence of flips, where in each step the most profitable (least unprofitable) flip is chosen from the flips that change the value of a variable that has not been flipped before in this sequence. An arbitrary rule may be used for breaking ties.

Lemma 6.2. POS NAE MAX-3SAT *with the Kernighan-Lin neighborhood function is PLS-complete.*

Proof. We prove the theorem by giving a PLS-reduction from MAX-CIRCUIT/flip, which by Theorem 6.4 is PLS-complete.

Preliminaries. The Boolean function $\text{nor}(a,b)$ is defined by $\neg a \wedge \neg b$, which means that $\text{nor}(a,b) = 1$ if and only if $a = b = 0$. It can be verified that $\neg a \equiv \text{nor}(a,0)$, $a \wedge b \equiv \text{nor}(\neg a, \neg b)$, and $a \vee b \equiv \neg \text{nor}(a,b)$. As a result, any Boolean circuit D consisting of gates from $\{\wedge, \vee, \neg\}$ can be transformed into a Boolean circuit D' consisting of only nor-gates, such that D' computes the same function as D and the sizes of D and D' differ by a constant factor. Therefore, we can assume in this proof that a Boolean circuit that defines a problem instance of MAX-CIRCUIT/flip consists of nor-gates only. We also assume that, when given an input vector $x \in \{0,1\}^n$, a Boolean circuit not only computes an output vector $y \in \{0,1\}^m$, but also an output vector $z \in \{0,1\}^n$ that equals the lexicographically first higher-cost flip neighbor of x if x is not locally optimal and equals x otherwise. This assumption is made without loss of generality because such a circuit can be derived in polynomial time from any Boolean circuit D in the following way. We first add n copies of D, where the ith copy evaluates D for the input vector x in which the ith bit has been flipped. By adding some logic, we next connect the copies, such that a Boolean circuit is obtained that satisfies the property described. The third assumption we make is that a Boolean circuit does not contain nodes with both indegree and outdegree zero, i.e., nodes that are both input node and output node. These nodes can safely be removed as they always have the value one in a locally optimal solution.

Definition of PLS-reduction. We now start defining the PLS-reduction (φ_1, φ_2) from MAX-CIRCUIT/flip to POS NAE MAX-3SAT/Kernighan-Lin. Let D be a Boolean circuit defining a problem instance of the former local search problem. We define $|D|$ as the size of D and $M = (3|D|)^{3|D|}$ as a large number depending on $|D|$. Furthermore, let g_j be gate number j, where the gates are numbered from 1 to $|D| - n$ in such a way that an input of a gate g_j is either an input node or a gate with a smaller number. Without loss of generality we assume that $|D| \geq 4$.

We define the problem instance $\varphi_1(D)$ of local search problem POS NAE MAX-3SAT/Kernighan-Lin as follows. For every input node x_i and gate g_j we introduce the binary variables x_i and g_j that represent their values. Furthermore, we also introduce two primed variables x'_{i1} and x'_{i2} for each variable x_i and one primed variable g'_j for each gate variable g_j. The primed variables represent the negation of the unprimed variables. To ensure that in 'good' solutions the values of the primed variables meet this interpretation, we let $\varphi_1(D)$ contain the clauses $\{x_i, x'_{i1}\}$, $\{x_i, x'_{i2}\}$, and $\{g_j, g'_j\}$ for all $1 \leq i \leq n$ and $1 \leq j \leq |D| - n$, each having weight M. For each nor-gate g_j, we also add the three clauses $\{g_j, a, b\}$, $\{g_j, a, 1\}$, and $\{g_j, b, 1\}$, where variables a and b correspond to the gates or input nodes that form the input of gate g_j in D. All three clauses have weight $\alpha_j = M/(3|D|)^{2j}$. It can be verified that these three clauses are all satisfied by some truth assignment if and only if the value of variable g_j equals $\text{nor}(a,b)$. Correspondingly, gate variable g_j is said to

be consistent if these three clauses are all satisfied and it is said to be inconsistent otherwise. The three clauses are called consistency clauses. Regarding the weights of the consistency clauses, it can be verified that

$$\alpha_j = (3|D|)^2 \alpha_{j+1} \tag{6.1}$$

holds. We also introduce a clause $\{p,0\}$ of weight $M - 1$.

If a truth assignment satisfies all clauses presented above, we say that it is feasible. By the definition of the clauses, this means that a truth assignment is feasible if and only if the value of each primed variable is given by the negation of its unprimed counterpart, the values of variables x_i and g_j define a valid computation of Boolean circuit D, and $p = 1$. We write t_x for the uniquely determined feasible truth assignment that is implied by the computation of D on input $x \in \{0,1\}^n$. Furthermore, we let R be the set containing all feasible truth assignments, i.e., $R = \{t_x | x \in \{0,1\}^n\}$.

To problem instance $\varphi_1(D)$ we also add a clause $\{y_k, 0\}$ with weight 2^{k+1} for each output node y_k in D. From these clauses we can derive the cost of a computation corresponding to a feasible truth assignment because we have $y_k = 1$ if the clause is satisfied and $y_k = 0$ if this is not the case. Finally, we introduce a clause $\{x'_{i1}, z_i\}$ with weight two for each output node z_i. If this clause is not satisfied in a feasible truth assignment t_x, then this means that flipping bit i in x results in a flip-neighbor of x with better cost than x for Boolean circuit D. Note that variables y_k and z_i are gate variables. This concludes the definition of problem instance $\varphi_1(D)$. An overview is given in Table 6.1. Algorithm φ_2 returns solution x on input t_x and an arbitrary solution, for instance 0, on any other input.

Proof strategy. What remains to be proved is that the presented algorithms φ_1 and φ_2 define a valid PLS-reduction. We partition the set of clauses into two subsets, C_1 and C_2. The former subset contains the clauses written as $\{y_k, 0\}$ and $\{x'_{i1}, z_i\}$, and C_2 contains the other clauses, i.e., the clauses that involve the feasibility of a truth assignment. The minimum weight of any clause in C_2 is given by at least $\alpha_{|D|-n} = (3|D|)^{|D|+2n}$. Furthermore, we have that the total weight of the clauses from C_1 is given by $w(C_1) = \sum_{k=1}^{m} 2^{k+1} + 2n \leq 2^{m+2} + 2n$. As $2^{m+2} + 2n < 2^{|D|+1} + 2^{|D|} = 4 \cdot 2^{|D|}$ and $|D| \geq 4$, by assumption, this yields

$$w(C_1) < 3^{|D|}. \tag{6.2}$$

Hence, the total weight of the clauses in C_1 is strictly smaller than the minimum weight of any clause in C_2. This gives rise to the following important observation.

Observation. The minimum cost over all feasible truth assignments is strictly larger than the maximum cost over all infeasible truth assignments.

Hence, in terms of the transition graph $T_{\varphi_1(D)}$ we have that a solution in R has only outgoing arcs to other solutions in R. To prove that (φ_1, φ_2) defines a valid PLS-reduction, we now proceed in a similar way as we did in the proof of Theorem 6.4. We show that the transition graph $T_{\varphi_1(D)}$ satisfies the following two properties, where algorithm $B(x)$ gives the lexicographically first neighbor of x with better

Table 6.1. Clauses in $\varphi_1(D)$ with their weight, where $M = (3|D|)^{3|D|}$.

For each input node x_i with $1 \leq i \leq n$

Clause	Weight				
$\{x_i, x'_{i1}\}$	$M = (3	D)^{3	D	}$
$\{x_i, x'_{i2}\}$	M				

For each nor-gate $g_j = \text{nor}(a,b)$ with $1 \leq j \leq |D| - n$

Clause	Weight		
$\{g_j, g'_j\}$	M		
$\{g_j, a, b\}$	$\alpha_j = \frac{M}{(3	D)^{2j}}$
$\{g_j, a, 1\}$	α_j		
$\{g_j, b, 1\}$	α_j		

For each output node y_k and z_i with $1 \leq j \leq |D| - n$ and $1 \leq i \leq n$

Clause	Weight
$\{y_k, 0\}$	2^{k+1}
$\{x'_{i1}, z_i\}$	2

Others

Clause	Weight
$\{p, 0\}$	$M - 1$

cost and where B^i means that B is applied i times in succession. Note that B satisfies the properties of algorithm B in Definition 6.3.

- *Condition 1.* If $x' = B(x)$, then $T_{\varphi_1(D)}$ contains the arc $(t_x, t_{x'})$. Conversely, if $T_{\varphi_1(D)}$ contains the arc $(t_x, t_{x'})$, then an i exists with $1 \leq i \leq n$, such that $x' = B^i(x)$. See Figure 6.5.
- *Condition 2.* For any truth assignment $t \notin R$, transition graph $T_{\varphi_1(D)}$ contains a path to a feasible truth assignment $t_x \in R$.

Proving these two conditions implies that (φ_1, φ_2) is a PLS-reduction. As in the proof of Theorem 6.4, the first condition may be weakened for the purpose of this result since we do not need the second part of the condition. However, besides providing further insight, it is used in Section 6.2.2, where we study the complexity of iterative improvement.

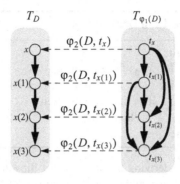

Figure 6.5. Possible correspondence between arcs in transition graph T_D and arcs in transition graph $T_{\varphi_1(D)}$, where $x(i) = B^i(x)$.

Proof of first part of Condition 1. It follows from the construction of $\varphi_1(D)$ that the higher the cost of $x \in \{0,1\}^n$ in D, the higher the cost of t_x in $\varphi_1(D)$. Hence, to prove the first part of Condition 1, it suffices to show that for any $x \in \{0,1\}^n$ and $x' = B(x)$ the truth assignment $t_{x'}$ is a neighbor of t_x. To prove this we derive the successive Kernighan-Lin neighbors of t_x. When deriving these neighbors, we will encounter $t_{x'}$, which proves that $t_{x'}$ is indeed a neighbor of t_x. More precisely, we show that the sequence of flips computed by the algorithm of Figure 6.6 specifies the successive flips that are performed by the Kernighan-Lin neighborhood function to determine the neighborhood of t_x. In the algorithm, i_0 denotes the unique bit in which x and x' differ. It can be verified that this sequence of flips does indeed result in $t_{x'}$.

We start with proving the validity of the first flip indicated in the algorithm of Figure 6.6, i.e., we prove that variable $x'_{i_0,1}$ is the first variable that is flipped. Flipping a gate variable g_j in t_x loses at least the clause $\{g_j, g'_j\}$ of weight M and a consistency clause. Furthermore, the only unsatisfied clauses are in C_1 and, as shown above, these clauses have a total weight lower than any single consistency clause. This implies that flipping g_j results in a cost reduction of at least M. Next, consider flipping a variable p, x_i, x'_{i1}, x'_{i2}, or g'_j. This loses at least one of the four clauses $\{p, 0\}, \{x_i, x'_{i1}\}, \{x_i, x'_{i2}\}$, and $\{g_j, g'_j\}$ of weight $M - 1$ or M. Furthermore, the only unsatisfied clause that can be satisfied as a result of such a flip is clause $\{x'_{i_0,1}, z_{i_0}\}$ with weight two. Note that $z_i = x_i$ for all $i \neq i_0$. Hence, a cost reduction of $M - 2$ is the best we can achieve and this loss can only be obtained by flipping $x'_{i_0,1}$. As a result, the first neighbor $t^{(1)}$ of t_x for $\varphi_1(D)$ is obtained by flipping variable $x'_{i_0,1}$, which is in accordance with the algorithm given in Figure 6.6

We next show that the second neighbor is obtained by flipping x_{i_0} in $t^{(1)}$. This flip gains weight M via $\{x_{i_0}, x'_{i_0,1}\}$, which is lost again via $\{x_{i_0}, x'_{i_0,2}\}$. Moreover, each gate g_j that has x_{i_0} as its input loses some of its three consistency clauses. The weight of one such clause is α_j and it is thus bounded from above by α_1. As a result,

algorithm Derive Neighbors

begin
 flip $x'_{i_0,1}$; flip x_{i_0}; flip $x'_{i_0,2}$;
 for $(j := 1; j \le |D| - n; j := j + 1)$ **do**
 if g_j inconsistent **then**
 flip g_j; flip g'_j;
end;

Figure 6.6. Algorithm for deriving the sequence of flips that are successively performed by the Kernighan-Lin neighborhood function to determine the neighborhood of t_x. The algorithm does not give the complete sequence of $|U|$ flips, but stops when it reaches neighbor $t_{x'}$ with $x' = B(x)$.

the total loss is bounded by the number of gates $|D| - n$ times the maximum loss $3\alpha_1$ per gate, which can easily be shown to be less than $M/(3|D|)$. Hence, flipping x_{i_0} reduces the cost of $t_x^{(1)}$ by strictly less than $M - 1$, while flipping any other variable still yields a reduction of at least $M - 1$. Hence, $t^{(2)}$ is obtained by flipping x_{i_0} in $t^{(1)}$. Note that in $t^{(2)}$ all primed variables, except for $x'_{i_0,2}$, are the complements of the unprimed ones and that one or more gate variables are inconsistent.

Once we have $t^{(2)}$, the best we can do is to satisfy clause $\{x_{i_0}, x'_{i_0,2}\}$ with weight M via flipping $x'_{i_0,2}$, which does not unsatisfy any satisfied clause. That this is indeed the best we can do can be seen as follows: except for $\{x_{i_0}, x'_{i_0,2}\}$, all other unsatisfied clauses have a total weight strictly less than M since, as shown above, the unsatisfied consistency clauses have a total weight of less than $M/(3|D|)$ and, by (6.2), the clauses from C_1 have a total weight $w(C_1)$ of at most $3^{|D|} < M/(3|D|)$. Hence, the third neighbor $t^{(3)}$ is obtained from $t^{(2)}$ by flipping $x'_{i_0,2}$. Note that in $t^{(3)}$ all primed variables have now become the complements of the unprimed ones. However, one or more gate variables are still inconsistent.

To prove the validity of the algorithm given in Figure 6.6, and thus the first part of Condition 1, we now have to show that in the next steps the values of the inconsistent gate variables are flipped in increasing order. We prove the claim by showing that if $p = 1$ and if all primed variables are the complements of the unprimed ones in a neighbor $t^{(l)}$ of t_x, then neighbor $t^{(l+1)}$ is obtained by flipping the first inconsistent gate variable g_j followed by its primed version g'_j.

Let a and b be the two inputs of gate g_j. As g_j is inconsistent, at least one of the three clauses $\{g_j, a, b\}$, $\{g_j, a, 1\}$, and $\{g_j, b, 1\}$ is not satisfied by $t^{(l)}$. To satisfy them all, it suffices to flip gate variable g_j. This results in a cost improvement of at least α_j. However, some of the $|D| - n - j$ variables corresponding to gates with a larger number may become inconsistent. This yields a loss of at most $3(|D| - n - j)\alpha_{j+1}$, which by (6.1) is strictly smaller than $(1/|D|)\alpha_j$. However, clause $\{g_j, g'_j\}$ also becomes unsatisfied, which gives an additional loss of M,

and clauses $\{x'_{i1}, z_i\}$ and $\{y_k, 0\}$ from set C_1 may result in a total loss of at most $w(C_1) < 3^{|D|} < (1/|D|)\alpha_j$. Hence, flipping g_j gives rise to a total loss of at most $M - \alpha_j(1 - 2/|D|)$. We now show that flipping g_j is the best alternative by showing that the cost reduction implied by flipping any other variable is strictly larger than this upper bound.

The claim is clearly true for the variables p, x'_{i1}, x'_{i2}, and g'_j as flipping any of them results in a cost reduction of at least $M - 2$. Suppose that we flip a variable x_i. In that case, we lose $2M$ via the two clauses $\{x_i, x'_{i1}\}$ and $\{x_i, x'_{i2}\}$. In the same way as done for $t^{(2)}$ it can be verified that the total sum of the weights of the unsatisfied clauses is less than M. Hence, by flipping x_i we cannot obtain a cost reduction of at most the derived upper bound for g_j.

Next, assume that we flip a gate $g_{j'}$ with $j' > j$. Then, at most $|D| - n - j'$ gates become consistent, which yields a gain of at most $3(|D| - n - j')\alpha_j < (1/|D|)\alpha_j$ via the consistency clauses in C_2. Furthermore, the unsatisfied clauses from C_1 may imply a gain of at most $w(C_1) < (1/|D|)\alpha_j$. Together with the loss of M that results from clause $\{g_{j'}, g'_{j'}\}$, we get a total loss of strictly more than $M - \alpha_j(2/|D|)$. Hence, because $|D| \geq 4$, by assumption, the cost reduction obtained by flipping g'_j is strictly larger than the derived upper bound $M - \alpha_j(1 - 2/|D|)$ on the cost reduction obtained by flipping g_j. This implies that flipping g_j is preferred to flipping any other gate variable $g_{j'}$ with $j' > j$.

As a last alternative to flipping g_j, we consider flipping a gate variable $g_{j'}$ with $j' < j$. In that case gate $g_{j'}$ becomes inconsistent, yielding a loss of at least $\alpha_{j'}$. However, it may be that all inconsistent gates in $t^{(l)}$ become consistent, giving a gain of at most $3(|D| - n - j)\alpha_j < (1/|D|)\alpha_{j'}$, and that all clauses from C_1 become satisfied, giving a gain of at most $w(C_1) < (1/|D|)\alpha_{j'}$. Combined with the loss of M via $\{g_{j'}, g'_{j'}\}$, we get a total loss of at least $M + \alpha_{j'}(1 - 2/|D|)$, which is obviously larger than the upper bound $M - \alpha_j(1 - 2/|D|)$.

This concludes the proof that gate g_j is flipped in truth assignment $t^{(l)}$. In much the same way as we showed that variable $x'_{i_0,2}$ is flipped in truth assignment $t^{(2)}$ it can be shown that a flip of g_j is directly succeeded by a flip of g'_j. This proves that if $x' = B(x)$, then the transition graph $T_{\varphi_1(D)}$ contains the arc $(t_x, t_{x'})$.

Proof of second part of Condition 1. Let $t_{x(1)}, t_{x(2)}, \ldots, t_{x(r)}$ be all the feasible truth assignments that are in the neighborhood of a feasible truth assignment t_x. Furthermore, let them be put in the order in which they are constructed when deriving the neighborhood of t_x. As each variable may only be flipped once when deriving the neighborhood of t_x, each variable x_i with $1 \leq i \leq n$ can only change value once. As a result, we have $r \leq n$. To complete the proof of the second part of Condition 1, we now show that $x(l) = B^l(x)$ holds for all $1 \leq l \leq r$.

Suppose that x is locally optimal for D, in which case $B(x)$ does not exist. Then $z_i = x_i$ for all i, which yields that flipping any variable, except for p, results in a cost reduction of at least M. Flipping variable p results in a cost reduction of only $M - 1$. As p may only be flipped once, this implies that all neighbors of t_x are infeasible, i.e., $r = 0$. Furthermore, it follows from the proof of the first part of Condition 1

that if $x(1) = B(x)$ exists, then the first neighbor of t_x from R is given by $t_{x(1)}$. Based on these observations, we can conclude that if, except for p, the blocking of each variable is removed after deriving a neighbor $x(l)$, which means that variables that have been flipped may be flipped again, then the neighbors of t_x from R are given by $x(l) = B^l(x)$ with $1 \leq l \leq L$ for some L. Furthermore, $B^L(x)$ is locally optimal. However, since blockings are not removed, it may be the case that the path from $x(l)$ to $x(l+1)$ described by the algorithm in Figure 6.6 cannot be taken as it may impose the flipping of a variable that has already been blocked. We prove our claim that $x(l) = B^l(x)$ for all $1 \leq l \leq r$ by showing that if such a blockade of a path occurs after, say, deriving neighbor $t_{x(l)}$, then all further neighbors of t_x are infeasible, i.e., $l = r$. Above, we have already discussed the case that $B(x)$ does not exist. Hence, we can assume that $t_{x(l)}$ exists.

Consider the first blockade we encounter after deriving neighbor $t_{x(l)}$. As variables x_i and x'_{i2} are only flipped directly after x'_{i1}, they cannot be involved in the first blockade. Suppose that the blocked variable that we want to flip is a variable $x'_{i_0,1}$. This means that we are deriving the first neighbor after arriving at feasible solution $t_{x(l)}$. Since we want to flip $x'_{i_0,1}$, clause $\{x'_{i_0,1}, z_{i_0}\}$ is not satisfied by $t_{x(l)}$, but all clauses $\{x'_{i1}, z_i\}$ with $i \neq i_0$ are. Consequently, only flipping variable p yields a reduction of less than M, namely $M - 1$. As indicated above, after flipping p no other feasible solutions are encountered.

By the same argument used for x_i and x'_{i2}, we have that a gate variable g'_j cannot be involved in the first blockade. Hence, to prove the second part of Condition 1, it now only remains to be shown that after encountering a blocked variable g_j we can only obtain infeasible solutions. Let the inputs of g_j be given by a and b. As g_j is blocked and as g_j is consistent in t_x, at least one of its inputs, say a, has changed its value before obtaining neighbor $t_{x(l)}$. Furthermore, since g_j has become inconsistent after obtaining neighbor $t_{x(l)}$, the other input b also has been blocked at the moment under consideration. As a result, the two inputs will not change value anymore, which means that g_j will remain inconsistent. Hence, $t_{x(l)}$ is the last feasible neighboring solution derived for t_x.

Proof of Condition 2. To conclude the proof of the lemma, we now only have to prove Condition 2, which states that for any infeasible solution a path in $T_{\varphi_1(D)}$ exists to a feasible solution. Assume that this is not the case, which means that a locally optimal solution \hat{t} exists that is infeasible. We show that this gives rise to a contradiction. If $p = 0$, then flipping p yields a cost improvement of $M - 1$. Hence, $p = 1$ holds in \hat{t}. Next, suppose that \hat{t} contains a primed variable x'_{i1}, x'_{i2}, or g'_j that is not the negation of its unprimed counterpart. Then flipping the primed variable results in a cost improvement of M via one of the clauses $\{x_i, x'_{i1}\}$, $\{x_i, x'_{i2}\}$, or $\{g_j, g'_j\}$. This is more than the possible loss of 2 resulting from losing the weight of clause $\{x'_{i1}, z_i\}$. Hence, in addition to $p = 1$, we may assume in \hat{t} that the primed variables are the negations of the unprimed ones. When proving the first part of Condition 1, we derived that if this is the case, then a sequence of neighbors of \hat{t} is constructed by each time flipping the first inconsistent gate variable g_j followed by

its primed version g'_j. This ultimately results in a feasible truth assignment. Hence, the neighborhood of \hat{t} contains a feasible solution and, as already observed, we know that the cost of any feasible solution is higher than the maximum cost over all infeasible solutions. This contradicts our assumption that \hat{t} is locally optimal. □

Knowing that POS NAE MAX-3SAT/Kernighan-Lin is PLS-complete, we are now able to prove that UGP/Kernighan-Lin is PLS-complete. Compared to proofs that CIRCUIT/flip and POS NAE MAX-3SAT/Kernighan-Lin are PLS-complete, this proof is significantly less involved, but it nevertheless illustrates a typical PLS-reduction.

Theorem 6.5. UGP *with the Kernighan-Lin neighborhood function is PLS-complete.*

Proof. Let the cut of a partition (V_1, V_2) be defined as the set of edges that have one endpoint in V_1 and one in V_2. In Section 2.5 we showed that UGP is equivalent to its maximization variant in which we are asked to partition the nodes of a graph into two sets V_1, V_2 of equal size, such that the weight $c(V_1, V_2)$ of the resulting cut is maximized instead of minimized. When we inspect the given relation between the two variants, it follows easily that the corresponding local search problems under the Kernighan-Lin neighborhood function are also equivalent. As a result, we can prove the theorem by giving a PLS-reduction (φ_1, φ_2) from POS NAE MAX-3SAT/Kernighan-Lin to the maximization version of UGP with the Kernighan-Lin neighborhood function, where we use Lemma 6.2, which states that the former problem is PLS-complete.

Definition PLS-reduction. Let I be an arbitrary problem instance of POS NAE MAX-3SAT/Kernighan-Lin and let x_1, x_2, \ldots, x_m be the binary variables of I. The problem instance $\varphi_1(I)$ is defined by the following graph $G = (V, E)$. Node set V contains two nodes x_i and x'_i for each variable x_i in I. The nodes x_i and x'_i represent the value of the variable x_i and its negation, respectively. To V we also add the nodes y and z, which represent the constant one, and the nodes y' and z', which represent the constant zero. This concludes the definition of V.

The edge set E is partitioned into the two subsets E_1 and E_2. Set E_1 contains all edges that connect an unprimed node x_i with its primed variant. Furthermore, it contains an edge from each of the two one-nodes y and z to each of the two zero-nodes y' and z'. All these $m + 4$ edges are assigned the same large weight of $W + 1$, where W is defined as the total weight of all clauses in I. The following two conditions are necessary and sufficient to guarantee that a cut contains all edges from E_1.

- Nodes y and z that represent the constant one are both assigned to the same subset and nodes y' and z' that represent the constant zero are both assigned to the other subset.

- The nodes x_i and x'_i are assigned to different subsets for all $1 \leq i \leq m$.

If these two conditions are satisfied by a partition, we say that the partition is feasible; see Figure 6.7 for an example. We can assume the following one-to-one

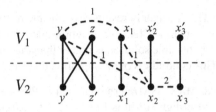

Figure 6.7. Feasible partition $P_t = (V_1, V_2)$ for the graph $\varphi_1(I)$, where I is the problem instance of POS NAE MAX-3SAT/Kernighan-Lin consisting of the two clauses $\{x_1, x_2, 1\}$ and $\{x_2, x_3\}$. Both clauses have a weight of two. The undashed edges are from E_1 and the dashed edges are from E_2. Each edge from E_1 has weight $W + 1 = 5$, and the weights of the edges from E_2 are denoted in the figure. Feasible partition P_t corresponds to truth assignment t with $x_1 = 1$ and $x_2 = x_3 = 0$. This truth assignment satisfies the clause $\{x_1, x_2, 1\}$ but not the clause $\{x_2, x_3\}$.

correspondence between a feasible partition and a truth assignment without violating the interpretation of the nodes x_i and x'_i as being the value of variable x_i and its negation. A variable x_i or its negation is assigned the value one if and only if the corresponding node is contained in the same subset as the variables y, z that represent the constant one. This means that if a variable or its negation is assigned the value zero, then the corresponding node is contained in the same subset as the two variables y' and z' that represent the constant zero. We define P_t as the feasible partition corresponding to truth assignment t and R as the set of all feasible partitions, i.e., $R = \{P_t \mid t \text{ truth assignment}\}$.

To be able to derive the cost of a truth assignment t from the feasible partition P_t, we add additional edges to E via E_2. For each clause $c = \{a, b\}$, we add the edge $\{a, b\}$ to E_2 with the same weight as clause c, where for the constants one and zero we use the nodes y and y', respectively. If such an edge contributes its weight to the cost of a feasible partition P_t, then a and b are assigned to different subsets, which implies that the clause $\{a, b\}$ is satisfied by truth assignment t.

For a clause $c = \{a, b, c\}$ of weight w, we include the three edges $\{a, b\}$, $\{b, c\}$, and $\{a, c\}$ in E_2, each with weight $w/2$. It can be verified that the cut of a potentially infeasible partition contains either two out of the three edges or none of the three edges. Moreover, for a feasible partition P_t the former case implies that clause c is satisfied by t and the latter case implies that c is not satisfied by t. As a result, the edges contribute w to the cost of a feasible partition P_t if c is satisfied by t and they do not contribute otherwise. We note that if a pair of nodes a, b occurs in several clauses, we assign to edge $\{a, b\}$ the weight obtained by summing up the weights that arise from all these clauses. This concludes the definition of G and therefore of problem instance $\varphi_1(I)$. To complete the definition of our PLS-reduction, we define algorithm φ_2 in such a way that it returns for a potentially infeasible partition P the truth assignment that assigns to a variable x_i the value one if and only if it is contained in the same subset as variable y. Note that this algorithm complies with the given relation between a truth assignment t and the feasible partition P_t.

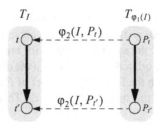

Figure 6.8. Correspondence between arc in transition graph T_I and arc in transition graph $T_{\varphi_1(I)}$.

Proof strategy. As the total weight W of the edges from E_2 that can be in the cut of a partition is strictly smaller than the weight $W + 1$ of a single edge in E_1 and as the cut of a feasible partition contains all edges from E_1, we can make a similar observation to that made in the proof of Lemma 6.2.

 Observation. The minimum cost over all feasible partitions is strictly larger than the maximum cost over all infeasible partitions.

This implies that, for any feasible partition P_t, the transition graph $T_{\varphi_1(I)}$ of $\varphi_1(I)$ only contains outgoing arcs to other feasible solutions. Consequently, to prove that (φ_1, φ_2) correctly defines a PLS-reduction, it suffices to prove that the transition graph satisfies the following two conditions.

- *Condition 1.* $T_{\varphi_1(I)}$ contains an arc $(P_t, P_{t'})$ if and only if T_I contains the arc (t, t'). See Figure 6.8.
- *Condition 2.* For any partition $P \notin R$, transition graph $T_{\varphi_1(I)}$ contains a path to a feasible partition.

In the proof we assume that, to determine the Kernighan-Lin neighborhood of a partition, an arbitrary rule is used for breaking ties in the case that several pairs of nodes qualify for being selected for a swap. Given this rule, we may choose the tie-breaking rule used to determine the Kernighan-Lin neighborhood of a truth assignment as we like because the PLS-completeness result for POS NAE MAX-3SAT/Kernighan-Lin holds regardless of this rule.

Proof of Condition 1. As the cost of a feasible partition P_t is given by $|E_1|(W + 1)$ plus the cost of truth assignment t, the first condition holds if we can prove that for any truth assignment t, partition P_t and t have the same neighborhood. More precisely, it suffices to show that $P_{t'}$ is a neighbor of P_t for $t' \neq t$ if and only if t' is a neighbor of t. Until all pairs of nodes x_i, x_i' have been swapped, and thus blocked, we only swap pairs x_i, x_i' of nodes when deriving the neighbors of P_t. This is true since swapping such a pair in a feasible partition results again in a high-quality, feasible partition, while swapping any other pair of nodes results in a low-quality, infeasible partition. After all pairs of nodes x_i, x_i' have been swapped, the nodes y, z and y', z' are swapped in two steps, where the intermediate partition is infeasible. This brings us back to partition P_t.

Swapping x_i and x_i' has the same effect on the cost of a feasible partition as flipping the value of variable x_i on the cost of the corresponding truth assignment. As a result, we obtain that, if the same rule is used for breaking ties in the derivation of the Kernighan-Lin neighborhood of t as is used in the derivation of the Kernighan-Lin neighborhood of P_t, then the feasible neighbors of P_t, excluding P_t, correspond to the truth assignments in the neighborhood of t. This proves the first claimed condition given above.

Proof of Condition 2. To prove the second condition, suppose that it does not hold, which means that an infeasible partition \hat{P} exists that is locally optimal. We show that this gives rise to a contradiction. As $\hat{P} = (V_1, V_2)$ is infeasible, either V_1 contains a pair u, u' of corresponding nodes or the two one-nodes y, z are split. First of all, consider the former case. Obviously, V_2 also contains a pair v, v' of corresponding nodes as $|V_1| = |V_2|$. Swapping u and v leads to a weight gain of at least $W + 1$ via the edges from E_1 and it leads to a weight reduction of at most W via the edges from E_2. This contradicts the assumption that \hat{P} is locally optimal. Next, assume that the two one-nodes are split, but that V_1 does not contain a pair u, u' of corresponding nodes. The two zero-nodes then also have to be in different subsets. Swapping a zero-node from one subset with a one-node in the other subset yields a cost improvement of $2(W + 1)$. Again, this contradicts the local optimality of \hat{P}. □

6.2 Time Complexity of Iterative Improvement

While the focus of the previous section was on the computational complexity of local search problems, we now study the time complexity of the iterative improvement algorithm when it is used to solve such a problem. A local search problem can be formulated as finding a node in a transition graph with outdegree zero. Iterative improvement is the obvious algorithm for tackling this problem. It traverses a transition graph, which is directed and acyclic, until it reaches a node without outgoing arcs. The precise path that the algorithm follows is determined by the pivoting rule. This rule specifies the neighboring solution that is selected if a solution has more than one neighbor with better cost. In terms of the transition graph, the pivoting rule prescribes which arc is to be selected in a given node if it has multiple outgoing arcs. Ideally, we would like to use a pivoting rule that takes the shortest path to a local optimum. However, computing this rule may be NP-hard, as, for instance, is the case for TSP with the 2-change neighborhood function.

As most of the neighborhood functions used in practice induce relatively small neighborhoods, which is reflected in our definition of the class PLS, it is generally easy to derive efficient implementations of commonly used (heuristic) pivoting rules as first improvement and best improvement. Hence, with regard to the time complexity of iterative improvement, we are particularly interested in the number of iterations required by iterative improvement. Obviously, the maximum number of iterations required by iterative improvement is bounded from above by the longest path that occurs in the transition graph. This upper bound holds regardless of the

pivoting rule used and the solution with which the algorithm starts. Whether or not this bound is tight depends on the pivoting rule applied. We therefore note that one should take account of the assumptions made about this rule when confronted with a claim about the time complexity of iterative improvement. With regard to the starting solution of iterative improvement, we generally do not want to make any assumptions. A reason for this is that because the quality of a starting solution generally has a major effect on the quality of the final solution, we want to have the opportunity to tune the algorithm that derives a starting solution to the application we are considering. Another reason is that, in order to make multiple runs of iterative improvement successful, we like to initialize each run with a solution that is quite different from the starting solutions used in the other runs; see Chapter 7.

Suppose that we apply iterative improvement to solve a problem Π_{LS} in PLS and let c be the cost of the solution with which we initialize the algorithm. Because the cost function is assumed to be integral and because in each iteration of iterative improvement the cost of the solution improves, the number of iterations the algorithm needs for reaching a local optimum is bounded by $|c - f^*|$, where f^* is the cost of an optimal solution. From this, it follows that if in the combinatorial optimization problem underlying Π_{LS} the cost of a solution is polynomially bounded in the input size, then iterative improvement can be implemented to run in polynomial time for Π_{LS}. An example is MINIMUM GRAPH COLORING with the Kempe chain neighborhood function where we define the cost of a coloring as the number of colors used in the coloring; see Section 2.4.

Although not satisfying this strong constraint, the cost functions of many combinatorial optimization problems satisfy the weaker property that the cost of a solution is polynomially bounded in the input size and the magnitude of the largest number that occurs in the problem instance. In these cases iterative improvement has a pseudo-polynomial running time. Examples include the other combinatorial optimization problems discussed in this book, such as TSP, machine scheduling problems, STG, and UGP. For TSP the cost of a solution is, for instance, bounded by n times the longest distance in the distance matrix. In the combinatorial optimization problem underlying CIRCUIT/flip, which is defined in Section 6.1.1, the cost function does not even satisfy this weaker property.

This section is split into two parts. In Section 6.2.1, we first derive a negative result and a positive result for iterative improvement in an ad hoc manner, i.e., without employing general ideas. For these two results to hold, assumptions are made about the pivoting rule used. In Section 6.2.2 we further exploit the theory of PLS-completeness to derive negative results that hold regardless of the pivoting rule used.

6.2.1 Ad Hoc Strategies

The following theorem states the negative result that iterative improvement does not have a polynomial running time for METRIC TSP with the 2-change neighborhood function.

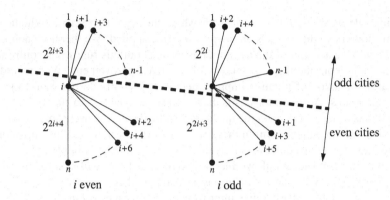

Figure 6.9. Definition of the distance matrix d in the problem instance of SYMMET-RIC TSP introduced in the proof of Theorem 6.6, where $i \geq 2$ is an arbitrary city that is either even (left) or odd (right).

Theorem 6.6. *For* METRIC TSP *with the 2-change neighborhood function, iterative improvement may require an exponential number of iterations to reach a local optimum if the pivoting rule is used that selects an improving move randomly.*

Proof. Consider a problem instance I of SYMMETRIC TSP instead of METRIC TSP. By adding to each entry d_{ij} of the distance matrix d the maximum distance d_{\max} occurring in d, we obtain a distance matrix that satisfies the triangle inequality. This modification does not affect a run of iterative improvement as for any tour it results in a length increase of exactly $n \cdot d_{\max}$. Hence, to prove the theorem it suffices to show that it holds for SYMMETRIC TSP.

We now proceed as follows. For any even n we construct a problem instance I of SYMMETRIC TSP consisting of n cities. For this problem we then derive a sequence of exponentially many tours that can be visited successively by the iterative improvement algorithm. Note that because iterative improvement runs in pseudo-polynomial time for SYMMETRIC TSP, the distance matrix of I will have to contain distances of exponential length.

Because we are considering SYMMETRIC TSP, we have $d_{ij} = d_{ji}$. Therefore, it suffices to define the distance d_{ij} for the case where $i \geq 2$ and either $j > i$ or $j = 1$. The definition of the distance matrix d is visualized in Figure 6.9. Formally, for even $i \geq 2$, the distance d_{ij} is given by 2^{2i+3} if j is odd and either $j = 1$ or $j > i$ and d_{ij} is given by $d_{ij} = 2^{2i+4}$ if j is even and $j > i$. For odd $i \geq 2$, we apply the same case distinction: d_{ij} is given by 2^{2i} if j is odd and either $j = 1$ or $j > i$ and it is given by 2^{2i+3} if j is even and $j > i$.

We prove that iterative improvement can take an exponential number of iterations to transform the tour $(1, 2, \ldots, n)$ into the tour $(1, n-1, n-2, \ldots, 2, n)$ for the problem instance described. We prove this result by showing by induction on q that the following assertion holds. Let tour τ contain a subtour $x, 2, 3, \ldots, 2q+1, y$

for a $q \leq \frac{n-2}{2}$, where x and y are cities, possibly 1 and $2q+2$, such that for all $2 \leq i \leq 2q+1$ we have $d_{x,i} = d_{1,i}$ and $d_{i,y} = d_{i,2q+2}$. Then a sequence of at least 2^{q-1} improving 2-changes exists that transforms tour τ into τ_q, where tour τ_q is equivalent to τ, except that the subtour $x, 2, 3, \ldots, 2q+1, y$ is replaced by the subtour $x, 2q+1, 2q, \ldots, 2, y$, in which the order of the $2q$ middle cities has been reversed.

We first prove the basis case $q = 1$. Tour $\tau_q = \tau_1$ that contains $x, 3, 2, y$ is obtained from tour τ that contains $x, 2, 3, y$ by the single 2-change that replaces the edges $\{x, 2\}$ and $\{3, y\}$ by $\{x, 3\}$ and $\{2, y\}$. As $2^{q-1} = 1$, the basis case now follows if τ_1 is shorter than τ. This is true because for the cost improvement $\Delta(\tau, \tau_1)$ of τ_1 in comparison with τ, we can derive

$$
\begin{aligned}
\Delta(\tau, \tau_1) &= d_{1,2} + d_{3,4} - d_{1,3} - d_{2,4} \\
&= 2^7 + 2^9 - 2^6 - 2^8 \\
&> 0.
\end{aligned}
$$

Next, consider the case $q \geq 2$. We show how τ can be transformed into τ_q in at least 2^{q-1} improving 2-changes, where we distinguish five phases in the transformation. The final tour in phase i is denoted by $\tau^{(i)}$. Hence, $\tau^{(5)} = \tau_q$.

In the first phase, $\tau^{(1)} = \tau_{q-1}$ is derived from τ. By the induction hypothesis this can be done in 2^{q-2} improving 2-changes. In Phase 2 we next change the order of the cities $2q$ and $2q+1$ in subtour $x, 2q-1, 2q-2, \ldots, 2, 2q, 2q+1, y$ of $\tau^{(1)}$ by performing a 2-change on the edges $\{2, 2q\}$ and $\{2q+1, y\}$. This results in a tour $\tau^{(2)}$ with shorter length than $\tau^{(1)}$ since

$$
\begin{aligned}
\Delta(\tau^{(1)}, \tau^{(2)}) &= d_{2,2q} + d_{2q+1,2q+2} - d_{2,2q+1} - d_{2q,2q+2} \\
&= 2^8 + 2^{4q+5} - 2^7 - 2^{4q+4} \\
&> 0.
\end{aligned}
$$

In Phase 3 the first and last edges $\{x, 2q-1\}$ and $\{2q, y\}$ occurring in subtour $x, 2q-1, 2q-2 \ldots, 2, 2q+1, 2q, y$ of $\tau^{(2)}$ are replaced by the two edges $\{x, 2q\}$ and $\{2q-1, y\}$. This yields a cost improvement of

$$
\begin{aligned}
\Delta(\tau^{(2)}, \tau^{(3)}) &= d_{1,2q-1} + d_{2q,2q+2} - d_{1,2q} - d_{2q-1,2q+2} \\
&= 2^{4q-2} + 2^{4q+4} - 2^{4q+3} - 2^{4q+1} \\
&> 0.
\end{aligned}
$$

The resulting tour $\tau^{(3)}$ contains $x, 2q, 2q+1, 2, 3, \ldots, 2q-1, y$. By the definition of the distance matrix, we have $d_{2q+1,i} = d_{1,i}$ and $d_{i,2q+2} = d_{i,2q}$ for all $2 \leq i \leq 2q$. As a result, we can apply the induction hypothesis to the subtour $x', 2, 3, \ldots, 2q-1, y'$ of $\tau^{(3)}$ with $x' = 2q+1$ and $y' = y$. This yields that $\tau^{(3)}$ can be transformed into $\tau^{(4)}$ by at least 2^{q-2} improving 2-changes, where $\tau^{(4)}$ is given by $\tau^{(3)}$ with the order of the cities $2, 3, \ldots, 2q-1$ reversed, i.e., $\tau^{(4)}$ contains $x, 2q, 2q+1, 2q-1, 2q-2, \ldots, 2, y$. In the final phase we reverse the order of the cities $2q$ and $2q+1$ by performing a 2-change on $\{x, 2q\}$ and $\{2q+1, 2q-1\}$. This again results in an improvement of

the tour as we have

$$
\begin{aligned}
\Delta(\tau^{(4)}, \tau^{(5)}) &= d_{1,2q} + d_{2q-1,2q+1} - d_{1,2q+1} - d_{2q-1,2q} \\
&= 2^{4q+3} + 2^{4q-2} - 2^{4q+2} - 2^{4q+1} \\
&> 0.
\end{aligned}
$$

In conclusion, we have transformed τ into $\tau^{(5)} = \tau_q$ by only performing improving 2-changes. The total number of 2-changes is at least 2^{q-2} in Phase 1, one in Phases 2 and 3, at least 2^{q-2} in Phase 4, and one in Phase 5. This makes a total of more than 2^{q-1} improving 2-changes. This proves the induction hypothesis and consequently it proves the theorem. □

We emphasize that the proof of the above theorem need not hold for alternative pivoting rules. By assuming, for instance, a pivoting rule that selects two edges that are closest to city one, we can transform the given tour $(1, 2, \ldots, n)$ into tour $(1, n-1, n-2, \ldots, 2, n)$ in one single move instead of in $2^{\frac{n-2}{2}}$ moves. A particularly interesting open question is whether iterative improvement may still require an exponential number of iterations when the frequently applied best-improvement pivoting rule is used.

Another interesting question is how many iterations iterative improvement requires when applied to EUCLIDEAN TSP with the 2-change neighborhood function. For this variant of TSP, Lemma 5.1 states that a locally optimal tour does not have edges that touch, intersect, or overlap. If the reverse were also true, then answering the question above would correspond to deriving the maximum number of cost-improving 2-changes required to reach a tour without any touching, intersecting, or overlapping edges. The reverse is, however, not true because the tour depicted in Figure 2.17 is not locally optimal despite the fact that it has this property. The following theorem nevertheless gives a bound on the number of cost-improving 2-changes that are required to obtain a tour with no edges that touch, intersect, or overlap. It focuses on the case that no three cities lie on a straight line, which implies that edges cannot touch or overlap.

Theorem 6.7. *Consider a problem instance of* EUCLIDEAN TSP *in which no three cities lie on a straight line. For an arbitrary tour, we have that repeatedly selecting any two intersecting edges for a 2-change yields a sequence of $\mathcal{O}(n^3)$ cost-improving 2-changes that transforms it into a tour without intersections.*

Proof. We define a function f from the set of tours to \mathbb{N} with the following two properties.

- The value $f(\tau)$ is bounded from above by n^3.
- If a tour τ contains intersecting edges, then performing a 2-change on any two intersecting edges yields, besides a shorter tour, a decrease of $f(\tau)$.

The existence of f obviously proves the theorem.

We use the term 'beam' to indicate an infinite straight line going through any two cities. Hence, as by assumption no three cities lie on a straight line, each edge lies on one unique beam. We define the index $f(e)$ of an edge e as the number of

beams that intersect, touch, or overlap e, and we define the index $f(\tau)$ of a tour τ as $\sum_{e \in \tau} f(e)$. We show that this function f satisfies the two properties given above, where we let τ denote an arbitrary tour.

The first property is easy. We have $\binom{n}{2} = \frac{n(n-1)}{2}$ edges and just as many beams. This gives $f(e) < n^2$ for any edge e and thus also $f(\tau) < n^3$.

Next consider the second property. Let e_1 and e_2 be any two intersecting edges in τ. Then performing a 2-change on e_1 and e_2 yields a shorter tour τ' in which these two edges are replaced by two new non-intersecting edges. We denote these new edges by e_1' and e_2'. The 2-change is visualized in Figure 5.2(a) for $e_1 = \{u_1, v_1\}$, $e_2 = \{u_2, v_2\}$, $e_1' = \{u_1, u_2\}$, and $e_2' = \{v_1, v_2\}$. The 2-change need not decrease the total number of intersections since new intersections may have been introduced. However, it does decrease $f(\tau)$, as shown in the following. It can be observed that the beams associated with e_1, e_2, e_1', and e_2' all contribute one to $f(e_1)$ and $f(e_2)$. However, while the beams associated with e_1 and e_2 contribute one to both $f(e_1')$ and $f(e_2')$, the beam associated with e_1' only contributes one to $f(e_1')$ and the beam associated with e_2' only contributes one to $f(e_2')$. Furthermore, for the beam associated with any other edge, it can be observed that if it intersects either e_1' or e_2' then it also intersects e_1 or e_2 and if it intersects both e_1' and e_2' then it also intersects both e_1 and e_2. From these observations, we can conclude that $f(\tau) - f(\tau') = f(e_1) + f(e_2) - f(e_1') - f(e_2') \geq 2$, which proves the second property. $\qquad\square$

In the remainder of this section we focus on the move neighborhood function for MULTIPROCESSOR SCHEDULING, as described in Section 2.2. Bear in mind that we assume use of the non-straightforward cost function that returns not only the makespan of a schedule, but also the number of critical machines it contains. Let the load l_i of a machine i be defined as in Section 5.2.2 and let $\Delta = l_{\max} - l_{\min}$, where l_{\max} and l_{\min} give the maximum and minimum load over all machines, respectively. Obviously, moving a job j from a machine i to a machine i' results in a better solution if and only if machine i is critical and $p_j < l_i - l_{i'}$. Note that if $p_j = l_i - l_{i'}$, then moving job j from i to i' implies that the new load of machine i' equals the old load of machine i and vice versa. It follows that for a critical machine i the set C_i of candidate jobs whose movement can realize a cost reduction is given by $C_i = \{j \mid j \in A_i \wedge p_j < \Delta\}$, where A_i contains the jobs assigned to machine i. We can actually achieve an improvement from a candidate job $j \in C_i$ by moving it to a machine with minimum load. We now propose the following pivoting rule: we select the move neighbor that is obtained by moving a candidate job $j \in C_i$ from a critical machine i to a machine with minimum load. Furthermore, we select job j from C_i such that it has maximal processing requirement, i.e., for each job $j' \in C_i$ with $p_{j'} > p_j$ we have $p_{j'} \geq \Delta$. The theorem below states that if we apply this pivoting rule then iterative improvement requires only $\mathcal{O}(nm)$ iterations.

The non-trivial pivoting rule we proposed can also be modeled in the neighborhood function. This can be done in the following way. Consider the neighborhood function obtained from the move neighborhood function by removing all arcs in the

move neighborhood graph that do not comply with the pivoting rule proposed. This neighborhood function combined with the simple pivoting rule that selects a random neighbor results in the same implementation of iterative improvement as the move neighborhood function combined with the non-trivial pivoting rule that we proposed. This remark generally holds, i.e., when specifying an iterative improvement implementation we can make a trade-off between assigning complexity to the neighborhood function and to the pivoting rule.

Theorem 6.8. *Consider* MULTIPROCESSOR SCHEDULING *with the move neighborhood function. Iterative improvement requires* $\mathcal{O}(nm)$ *iterations if a pivoting rule is used that selects a job with the largest possible processing requirement on a critical machine to move to a machine with minimum load.*

Proof. In the notation used in this proof we add superscript (k) to indicate that we are referring to the situation at the start of the kth iteration of iterative improvement. It can be verified that, as a function of k, $l_{\min}^{(k)}$ is non-decreasing, $l_{\max}^{(k)}$ is non-increasing, and, consequently, $\Delta^{(k)}$ is also non-increasing.

We show that once a job has been removed from a machine it never returns to that same machine. This proves the theorem as it implies that each of the n jobs can move at most $m-1$ times. Assume that the claim does not hold, which means that a job j exists that is removed from a machine i in iteration k_{leave} and is moved back onto machine i in iteration k_{return}. We show that this gives rise to a contradiction. We define k' as the first iteration after k_{leave} at the start of which machine i has the smallest load over all m machines. Furthermore, we let k be the last iteration before iteration k' in which a job has been removed from machine i. We denote this job by j_{last}. Note that $k_{\text{leave}} \leq k < k' \leq k_{\text{return}}$, $l_{\max}^{(k)} = l_i^{(k)}$, and $l_{\min}^{(k')} = l_i^{(k')}$.

As the maximum load l_{\max} is non-increasing, we obtain $l_{\max}^{(k')} \leq l_{\max}^{(k)}$ and thus $l_{\max}^{(k')} \leq l_i^{(k)}$. As a result, we derive

$$\Delta^{(k')} = l_{\max}^{(k')} - l_{\min}^{(k')} \leq l_i^{(k)} - l_i^{(k')}. \tag{6.3}$$

Between iterations k and k' no jobs are removed from or moved onto machine i. This implies $l_i^{(k')} = l_i^{(k+1)} = l_i^{(k)} - p_{j_{\text{last}}}$. As a result, the right-hand side of (6.3) is at most $p_{j_{\text{last}}}$, which gives $\Delta^{(k')} \leq p_{j_{\text{last}}}$.

As job j_{last} is removed from machine i in iteration k, we have $\Delta^{(k)} > p_{j_{\text{last}}}$. Since Δ is non-increasing this implies that $\Delta^{(k_{\text{leave}})} > p_{j_{\text{last}}}$ also holds. Furthermore, because by the minimality of k' no jobs are moved onto machine i between iterations k_{leave} and k', job j_{last} has been assigned to machine i at the start of iteration k_{leave}. As the assumed pivoting rule removes a job with the largest possible processing requirement from a critical machine, these two observations yield $p_{j_{\text{last}}} \leq p_j$. By $k' \leq k_{\text{return}}$ and $\Delta^{(k')} \leq p_{j_{\text{last}}}$, this implies that $\Delta^{(k_{\text{return}})} \leq p_j$, which contradicts our assumption that job j is removed from machine i in iteration k_{return}. This proves the theorem. □

6.2.2 General Strategy

In the previous section we proved Theorem 6.6, which states the negative result that if iterative improvement with the 2-change neighborhood function is applied to METRIC TSP, then the worst-case number of iterations required by the algorithm before reaching a local optimum cannot be bounded by a polynomial. In this result it is assumed that the pivoting rule is applied in which a lower-cost neighbor is chosen randomly. A question that arises is whether this bad worst-case behavior can be avoided by choosing a less naive pivoting rule or whether it is unavoidable due to the structure of the transition graph. This means that we are interested in the question of whether iterative improvement would still run in worst-case exponential time if for the pivoting rule we were to have an oracle at our disposal that gives us a neighbor that leads to a nearest local optimum. This problem, called the transition graph complexity problem, is formalized as follows. For the definition of the potential of a transition graph we refer to Definition 1.11.

Definition 6.9. For a given local search problem Π_{LS}, the *transition graph complexity problem* corresponds to deciding whether Π_{LS} can induce transition graphs with an exponentially large potential. □

In this section we study the transition graph complexity problem for PLS-complete problems. For PLS-complete problems we already know that, unless PO = PLS, they cannot be solved in polynomial time. Solving the transition graph complexity problem gives us additional information on whether the unavoidable exponential running time is caused by the structure of the transition graph or is only caused by the intractability of computing an optimal pivoting rule.

The reason we introduced local search problems is that they are the problems that are actually solved by iterative improvement. Solving a local search problem is generally not a goal in itself. Therefore, considerably less attention has been paid by scientists to solving these problems efficiently than to finding efficient algorithms for NP-hard combinatorial optimization problems and NP-complete decision problems. As a result, we have stronger evidence for the inequality P \neq NP, which by Theorem B.2 is equivalent to PO \neq NPO, than for PO \neq PLS. Note that, as discussed at the beginning of Section 6.1, the former inequality is also theoretically stronger than the latter one. This brings us to the second relevant implication of knowing that the potential of a transition graph can be exponentially large for a given PLS-complete problem. It proves an exponential worst-case running time for iterative improvement even in the case that PO = PLS. We nevertheless emphasize that it is commonly believed that PO \neq PLS holds.

To solve the transition graph complexity problem we proceed as follows. We defined PLS-reductions such that if $\Pi_{LS} \propto \Pi'_{LS}$ holds and if Π_{LS} cannot be solved in polynomial time, then Π'_{LS} cannot be solved in polynomial time either. Below we refine the definition of a PLS-reduction such that, in addition to this behavior, we have that if $\Pi_{LS} \propto \Pi'_{LS}$ holds and if a transition graph for Π_{LS} can have an exponentially large potential, then a transition graph of Π'_{LS} can also have an exponentially large potential. This more powerful PLS-reduction is called a tight PLS-reduction.

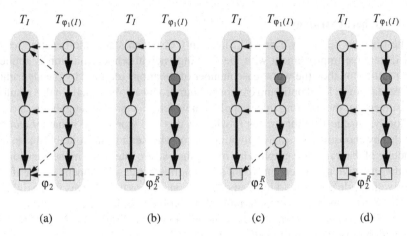

Figure 6.10. (a) Transition graphs T_I of $I = (S, f, N)$ and $T_{\varphi_1(I)}$ of $\varphi_1(I) = (S', f', N')$ and algorithm φ_2 for some local search problem instance I and PLS-reduction (φ_1, φ_2). A dashed arc originates from an $s' \in S'$ and points to $s = \varphi_2(I, s')$. (b)-(c) Invalid choices of R, where R is given by S' minus the shaded solutions in transition graph $T_{\varphi_1(I)}$. φ_2^R denotes algorithm φ_2 for I with the solution set S' of $\varphi_1(I)$ restricted to R. (d) Valid choice of R.

We show that the class PLS contains a local search problem for which the transition graph can have an exponentially large potential and we indicate that all PLS-reductions given in Section 6.1 are tight. This proves that all three problems for which we proved PLS-completeness can have transition graphs with exponentially large potential. Hence, for these problems iterative improvement needs in the worst case an exponential number of iterations to reach a local optimum even if a pivoting rule is used that takes the shortest path to a local optimum. This completes the outline of our approach for tackling the transition graph complexity problem. We now fill in the details.

Let (φ_1, φ_2) define a PLS-reduction between two local search problems Π_{LS} and Π'_{LS} and let $I = (S, f, N)$ be an arbitrary problem instance of Π_{LS} with image $\varphi_1(I) = (S', f', N')$. By definition, algorithm φ_2 maps any solution in S' to a solution in S, such that a local optimum in S' is also a local optimum in S. Figure 6.10(a) depicts possible transition graphs T_I and $T_{\varphi_1(I)}$ of I and $\varphi_1(I)$, respectively, and a possible choice for φ_2. For (φ_1, φ_2) to be a tight PLS-reduction, a subset R of S' has to exist that satisfies the following three conditions. In the first place, φ_2 has to be surjective for I when restricting the solution set S' of $\varphi_1(I)$ to R. This means that for each solution $s \in S$, a solution $s' \in R$ must exist that is mapped to s by algorithm φ_2. Furthermore, solution s' must be computable in polynomial time. Note that for (φ_1, φ_2) to be a standard PLS-reduction it is not necessary that φ_2 is surjective for a given problem instance, as follows from Figure 6.1. The second condition on R is that it contains all local optima of $\varphi_1(I)$. Figures 6.10(b) and 6.10(c) give

examples of R that do not satisfy the first and second condition, respectively, and Figure 6.10(d) gives an example of R that satisfies both conditions. The third and last condition that has to be satisfied is that if in $T_{\varphi_1(I)}$ a solution $s'_2 \in R$ can be reached from a solution $s'_1 \in R$ without visiting any other solution from R, then the distance between the corresponding solutions in S is not too large, i.e., it is polynomially bounded.

Definition 6.10. Local search problem Π_{LS} is *tightly PLS-reducible* to local search problem Π'_{LS}, denoted by $\Pi_{\mathrm{LS}} \propto_{\mathrm{tight}} \Pi'_{\mathrm{LS}}$, if a PLS-reduction (φ_1, φ_2) exists, such that for a polynomial p the following holds. For any problem instance $I = (S, f, N)$ of Π_{LS} with image instance $\varphi_1(I) = (S', f', N')$ of Π'_{LS}, we can choose a subset $R \subseteq S'$ that satisfies the following properties.

- For any $s \in S$ we can construct in polynomial time a solution $s' \in R$ with $\varphi_2(I, s') = s$.

- R contains all local optima in S'.

- If transition graph $T_{\varphi_1(I)}$ of $\varphi_1(I)$ contains a direct path from $s'_1 \in R$ to $s'_2 \in R$, then the distance from $s_1 = \varphi_2(I, s'_1)$ to $s_2 = \varphi_2(I, s'_2)$ in transition graph T_I of I is bounded from above by $p(|I|)$, where a path in $T_{\varphi_1(I)}$ is said to be direct if, except for the first and last solution, it only contains solutions outside R.

The pair (φ_1, φ_2) is called a *tight PLS-reduction*. □

Definition 6.11. A local search problem in PLS is *tightly PLS-complete* if each problem $\Pi_{\mathrm{LS}} \in$ PLS is tightly PLS-reducible to it. □

This definition of a tight PLS-reduction preserves the property that a transition graph induced by a problem can be exponential, as shown in the following. By the last condition in Definition 6.10, we have that if the distance between any two solutions $s'_1 \in R$ and $s'_2 \in R$ in $T_{\varphi_1(I)}$ is $d_{s'_1, s'_2}$, then $d_{s'_1, s'_2} \cdot p(|I|)$ gives an upper bound on the distance between the corresponding solutions $s_1 = \varphi_2(I, s'_1)$ and $s_2 = \varphi_2(I, s'_2)$ in T_I. As R contains all local optima, this implies that if $s' \in R$ is within k steps of a local optimum \hat{s}' in $T_{\varphi_1(I)}$, then solution $s = \varphi_2(I, s')$ is within $k \cdot p(|I|)$ steps of $\hat{s} = \varphi_2(I, \hat{s}')$, which by the definition of a PLS-reduction is a local optimum in S. Hence, by the second condition in Definition 6.10, the potential of T_I is at most $p(|I|)$ larger than the potential of $T_{\varphi_1(I)}$. Because the input size of $\varphi_1(I)$ is bounded from above by a polynomial in the input size of I, this implies that if T_I has an exponentially large potential, then $T_{\varphi_1(I)}$ must have one as well. We thus arrive at the following result.

Lemma 6.3. *If (φ_1, φ_2) defines a tight PLS-reduction from local search problem Π_{LS} to local search problem Π'_{LS} and if the potential of Π_{LS} cannot be bounded by a polynomial, then neither can the potential of Π'_{LS}.* □

We note that in order for this lemma to hold, the polynomial-time requirement in the first condition of Definition 6.10 is not needed. We nevertheless impose the requirement for a reason that can be found in the bibliographical notes at the end

of this chapter. By Lemma 6.3, it now suffices to indicate one local search problem in PLS for which the transition graph can have an exponentially large potential to conclude that this result holds for any tightly PLS-complete problem.

Lemma 6.4. *PLS contains a local search problem for which the potential of the transition graph cannot be bounded by a polynomial.*

Proof. Consider the trivial combinatorial optimization problem that, given an integer n, asks for the smallest integer i between 1 and n. This means that we want to minimize the cost function $f(i) = i$ with $1 \leq i \leq n$. For this problem, we consider the solution space consisting of the integers $1, 2, \ldots, n$ and we assume a neighborhood function in which integer 1 has no neighbors and integer $i \geq 2$ has only one neighbor, namely $i - 1$. This yields a local search problem that is obviously in PLS. Furthermore, the transition graph is given by a single directed path that successively visits the solutions $n, n - 1, \ldots, 1$. The potential of this graph equals the distance from n to 1, which is $n - 1$. This is exponentially large as the input size is only $\log n$. □

Theorem 6.9. *For a tightly PLS-complete problem, iterative improvement requires an exponential number of iterations in the worst case to reach a local optimum, regardless of the pivoting rule used.* □

Theorem 6.10. *The local search problems* MIN-CIRCUIT/*flip,* MAX-CIRCUIT/*flip,* POS NAE MAX-3SAT/*Kernighan-Lin, and* UGP/*Kernighan-Lin are all tightly PLS-complete.*

Proof. In Theorem 6.4 we prove that MIN-CIRCUIT/flip and MAX-CIRCUIT/flip are PLS-complete by means of two PLS-reductions in Lemma 6.1 and two in Theorem 6.4. Both PLS-reductions in Lemma 6.1 and the first one in Theorem 6.4 are obviously tight as they preserve solutions, costs, and neighborhoods. The same holds for the PLS-reduction that is mentioned implicitly at the beginning of Theorem 6.5 and that reduces UGP/Kernighan-Lin to this local search problem with UGP replaced by its maximization variant. We now show that the second PLS-reduction in Theorem 6.4 and the PLS-reduction in Theorem 6.5 are also tight.

For any problem instance I of the problem from which we reduced, we identified a set $R \subset S'$ with $(S', f') = \varphi_1(I)$ and proved the following two properties.

- $T_{\varphi_1(I)}$ contains a direct path from $s_1' \in R$ to $s_2' \in R$ if and only if T_I contains an arc (s_1, s_2) with $s_1 = \varphi_2(I, s_1')$, $s_2 = \varphi_2(I, s_2')$, and, in the case of CIRCUIT/flip, $s_2' = B(s_1')$, where B is computable in polynomial time.

- For any solution $s_1' \notin R$, transition graph $T_{\varphi_1(I)}$ contains a path to a solution $s_2' \in R$.

Furthermore, we indicated that φ_2 defines a bijective function from R to S. For the purpose of proving PLS-completeness, we did not need the only-if part of the first condition. However, by proving it, we showed that the PLS-reductions presented are tight. Note that the second condition implies that R contains all local optima in S'.

Next, consider the PLS-completeness proof of the local search problem POS NAE MAX-3SAT/Kernighan-Lin given in Lemma 6.2. In this proof we also identified for any problem instance I of the problem from which we reduced, in this case MAX-CIRCUIT/flip, a set $R \subset S'$ with $(S', f') = \varphi_1(I)$ such that the second property mentioned above holds and φ_2 defines a bijective function from R to S. The first property, however, has been replaced by the following one.

- If $s_2 = B(s_1)$ holds, where B is a polynomial-time algorithm, then $T_{\varphi_1(I)}$ contains the arc (s'_1, s'_2) with $s_1 = \varphi_2(I, s'_1)$ and $s_2 = \varphi_2(I, s'_2)$. Conversely, if $T_{\varphi_1(I)}$ contains a direct path from s'_1 to s'_2, then an i exists with $1 \le i \le n$, such that $s_2 = B^i(s_1)$.

Nevertheless, because $n \le |I|$ we can still conclude that the proposed PLS-reduction is tight. □

Suppose that for a combinatorial optimization problem Π we have two neighborhood functions N and N', where N' is a generalization of N, i.e., $N(s) \subseteq N'(s)$ for any solution s. For TSP, the 2-change and 3-change neighborhood functions are examples of N and N', respectively. Then the local search problem Π_{LS} specified by Π and N is PLS-complete if this is the case for the local search problem Π'_{LS} specified by Π and N'. More generally, this means that to prove PLS-completeness results for one combinatorial optimization problem and multiple neighborhood functions it suffices to prove that the weakest neighborhood function induces a PLS-complete problem. However, if we want to know whether the induced local search problems are *tightly* PLS-complete, we can no longer restrict ourselves to the weakest neighborhood function. It may be the case that Π_{LS} is tightly PLS-complete, while this is not the case for Π'_{LS}. The reason for this is that the generalization can add arcs to the transition graph, which may decrease its potential to a polynomially bounded value.

We refer you to Appendix C for a list of local search problems that have been proven to be (tightly) PLS-complete.

6.3 Bibliographical Notes

The complexity theory for local search problems presented in Section 6.1 has been introduced by Johnson, Papadimitriou & Yannakakis [1988]. They have also proved that CIRCUIT/flip and UGP/Kernighan-Lin are PLS-complete. However, our presentation follows more closely the paper of Yannakakis [1997]. In Appendix C we give a list of PLS-complete local search problems and references to the papers in which the PLS-completeness has been proved.

With regard to TSP with the 2-change neighborhood function, we mentioned in Section 6.2 that it is NP-hard to compute an optimal pivoting rule for iterative improvement or, more precisely, to determine the shortest path for a problem instance from a solution to a locally optimal solution in the corresponding transition graph. This is proved by Fischer [1995], who shows that it is NP-complete to decide for a problem instance I, a tour τ, and an integer k given in unary whether the transition graph of I contains a path of length at most k from τ to a locally optimal tour.

We also indicate in Section 6.2 that the applied pivoting rule can have a substantial effect on the number of iterations required by iterative improvement before a local optimum is reached. To substantiate this claim, consider the problem in which we are asked to find a maximum flow from a given source node to a given target node in a directed graph. An exact neighborhood function for this problem is the one in which a flow f' is a neighbor of flow f if it can be obtained by augmenting the flow along a path by as large a flow as possible. Papadimitriou & Steiglitz [1982] show that if we select an augmenting path arbitrarily, then in the worst case iterative improvement requires an exponential number of iterations, while only a polynomial number of iterations are required when the pivoting rule always selects a shortest augmenting path.

The proof that iterative improvement may require an exponential number of iterations in the worst case for METRIC TSP with the 2-change neighborhood function has been provided by Luecker [1976]. Although this result is often referred to, the proof has never been published. In Section 2.1 we defined the k-change neighborhood of a tour as the set of tours that can be obtained by replacing at most k edges. Alternatively, this neighborhood is also defined in the literature as the set of tours that can be obtained by replacing exactly k edges. The result of Luecker [1976] trivially translates to any $k \geq 2$ for our definition of k-change. Chandra, Karloff & Tovey [1999] generalize the result of Luecker [1976] to any $k \geq 2$ for the latter alternative definition of k-change. Their proof is based on the proof of Luecker [1976] given in Section 6.2.1.

The question of whether we can implement 2-Opt in such a way that it only needs a polynomial number of iterations to reach a local optimum is still unanswered. Nevertheless, for EUCLIDEAN TSP Chandra, Karloff & Tovey [1999] prove that if the n cities are uniformly distributed points in the unit square, then the expected number of iterations required by 2-Opt is $\mathcal{O}(n^{10} \log n)$, regardless of the pivoting rule that is used. Punnen, Margot & Kabadi [2003] prove for SYMMETRIC TSP that if we apply the best-improvement pivoting rule, then 2-Opt finds in polynomial time a tour τ that is no worse than the average cost of any tour, even though τ need not be locally optimal.

Theorem 6.7 is due to Van Leeuwen & Schoone [1981]. They prove that in $\mathcal{O}(n^3)$ steps a tour can be transformed into a tour that does not contain edges that touch, intersect, or overlap, where in each step the tour length is decreased. Unlike Theorem 6.7, they allow multiple cities to lie on a straight line, but they do not require that in each step a 2-change is performed.

For MULTIPROCESSOR SCHEDULING with the move neighborhood function, Brucker, Hurink & Werner [1997] prove that, if we assume a pivoting rule that allows any job from a critical machine i to move to a machine with minimum load and not, as in Theorem 6.8, only a job from i with largest possible processing requirement, then iterative improvement requires $\mathcal{O}(n^2)$ iterations to reach a local optimum (see also Exercise 1). Hurkens & Vredeveld [2003] and Vredeveld [2002] show that this bound cannot be improved for $m = 2$ because in that case the algorithm may need $\Omega(n^2)$ iterations when using such a pivoting rule. This result can easily be generalized to any $m \geq 2$.

For a similar pivoting rule, Vredeveld [2002] shows that $Q||C_{max}$ with the move neighborhood function is guaranteed to find a local optimum in $\mathcal{O}(n^2m)$ iterations. Hurkens & Vredeveld [2003] and Vredeveld [2002] derive that if the pivoting rule only selects a job from a critical machine with maximal possible processing requirement, then the bound on the number of required iterations improves to $\mathcal{O}(nm)$. We proved Theorem 6.8 in a similar way.

In Section 6.2.2 we introduced tight PLS-reductions for proving that a local search problem may induce transition graphs with an exponentially large potential. The notion of tight PLS-reductions has been introduced by Schäffer & Yannakakis [1991]. We used a slightly more general definition in the sense that Schäffer & Yannakakis [1991] require for the third condition in Definition 6.10 that the distance from s_1 to s_2 is either zero or one. We use our generalization in the proof of Theorem 6.10.

Tight PLS-reductions are also used to derive the complexity of the standard local optimum problem. This is the problem of determining for a given initial solution a local optimum that can be derived from it by iterative improvement. In other words, we are given a local search problem instance $I = (S, f, N)$ and an initial solution $s \in S$ and we are asked to find a sink of the transition graph of I that is reachable from s. Papadimitriou, Schäffer & Yannakakis [1990] prove that the standard local optimum version of a tightly PLS-complete problem is PSPACE-complete, which is an even stronger indication of intractability than NP-completeness [Garey & Johnson, 1979]. In Section 6.2.2 we mentioned that the polynomial-time requirement in the first condition of Definition 6.10 of a tight PLS-reduction is not needed to prove that local search problems may induce exponentially large transition graphs. The requirement is, however, used to prove this PSPACE-completeness result.

From the tight PLS-completeness of STABLE CONFIGURATION [Schäffer & Yannakakis, 1991], as defined in Appendix C, it follows that iterative improvement with the flip neighborhood function may require an exponential number of iterations in order to arrive at a stable configuration for a Hopfield network, regardless of the pivoting rule used. This result has already been proved by construction by Haken [1989].

Finally, we refer to Tovey [1997] for a discussion of worst-case and average-case results on the time complexity of local search for well-structured neighborhood graphs, such as hypercubes, that are, however, not implied by specific combinatorial optimization problems.

6.4 Exercises

1. Theorem 6.8 states that iterative improvement runs in $\mathcal{O}(nm)$ iterations for MUL-
 TIPROCESSOR SCHEDULING with the move neighborhood function if the pivot-
 ing rule is used that selects a job with the largest possible processing requirement
 on a critical machine to be moved to a machine with minimum load. Suppose
 that we use the alternative pivoting rule in which an arbitrary job from a critical
 machine is selected to be moved to a machine with minimum load.

a) Prove that between two moves of the same job j, at least one job (possibly job j) has been moved for the last time.

b) Use the result of (a) to derive an upper bound on the number of iterations required by iterative improvement.

2. Consider the exact neighborhood function for MINIMUM SPANNING TREE given in Theorem 5.1, i.e., consider the neighborhood function in which a spanning tree T' is a neighbor of a spanning tree T if and only if it can be obtained by first adding an arbitrary edge to T, which produces a cycle c, and by next deleting an arbitrary edge from cycle c.
Prove that if iterative improvement uses this neighborhood function, it finds a globally optimal solution in $\mathcal{O}(nm)$ iterations for a graph with n nodes and m edges.

3. Prove the following statements, where we refer you to Appendix C for the definitions of the local search problems.

a) POS NAE MAX-3SAT/flip \propto_{tight} MAX-CUT/flip.

b) MAX-CUT/flip \propto_{tight} STABLE CONFIGURATION.

c) MAX-CUT/flip \propto_{tight} UGP/swap.

d) MAX-CUT/flip \propto_{tight} MAX-2SAT/flip.

e) POS NAE MAX-3SAT/flip \propto_{tight} PURE NASH EQUILIBRIA IN CONGESTION GAMES.

7

Metaheuristics

Despite the complexity results presented in Chapter 6, iterative improvement is often able to compute a locally optimal solution relatively quickly. Furthermore, the quality of such a solution is generally good. However, only a small fraction of the local optima are usually really close to optimality and the worst local optima may be of a relatively poor quality. One way to obtain higher-quality solutions is to resort to a more powerful neighborhood function, although this may involve more computation time. This can, for instance, be achieved by applying variable-depth search, which we discussed in Section 2.5, or, in the case of k-Opt for TSP, by choosing a larger value of k. An alternative approach is not to use iterative improvement, but a metaheuristic that, unlike iterative improvement, does not stop at the first local optimum it encounters. Several such metaheuristics have been introduced. They can be classified into two categories: those that perform a single long walk in the neighborhood graph, thereby using some kind of mechanism to escape from local optima, and those that perform multiple walks in the neighborhood graph.

Representatives of the first strategy are characterized by their strategy for performing non-improving moves. This is required to prevent them from stopping at a local optimum. The most popular metaheuristics falling into this category are *simulated annealing* and *tabu search*. These two algorithms are discussed in Sections 7.1 and 7.2. Simulated annealing is a randomized algorithm that accepts any deterioration with a positive probability, while tabu search is in its basic form deterministic and enforces a move to a neighbor that leads to the smallest cost deterioration in the

case that a cost-improving neighbor does not exist. Tabu search uses a memory to keep track of the solutions it visited recently to avoid revisiting them.

In Sections 7.3 and 7.4 we present metaheuristics that prevent stopping at a local optimum by performing multiple walks in the neighborhood graph.

We do not aim to give a complete overview of the local search metaheuristics that are used, but restrict ourselves to presenting the basic versions of the most popular ones. We also do not elaborate on practical guidelines to apply the metaheuristics discussed. However, we stress that a successful application of a metaheuristic depends heavily on how well it is tuned to the problem at hand. References to more detailed treatments of the metaheuristics are given in the bibliographical notes at the end of this chapter.

We adopt the same notational conventions as in Section 1.1. This means that S and f together define a problem instance of an arbitrary combinatorial optimization problem and that N gives a neighborhood function for S. To simplify the presentation we restrict ourselves without loss of generality to minimization problems.

7.1 Simulated Annealing

Consider iterative improvement with the first-improvement pivoting rule, where a neighbor of a solution s is selected uniformly at random from its neighborhood $N(s)$. Hence, at a solution s, each neighbor $s' \in N(s)$ is generated with a probability of $1/|N(s)|$ and the first generated solution s' that has better cost than s is accepted, where accepting solution s' means that it replaces s as the current solution. The algorithm stops if it arrives at a local optimum.

A naive modification to this iterative improvement implementation to allow non-improving moves is to simply accept each suggested move, regardless of the resulting change in cost. However, the resulting algorithm, which is called *random search*, performs a random walk through the neighborhood graph and will consequently show a poor performance as the fraction of good solutions in the exponentially large solution space is generally small.

A possible improvement over random search is to accept a move from solution s to s' if and only if the increase in cost is strictly smaller than a threshold t, which is formalized by $f(s') - f(s) < t$. The question now is how to choose a suitable threshold. On the one hand, we prefer a large threshold to enable us to explore different parts of the solution space; but, on the other hand, a small threshold has the positive effect of guiding the search to areas in the solution space where the good solutions can be found. To enable both qualities we introduce the possibility of letting the threshold depend on the iteration in which it is inspected, i.e., each iteration k has its own threshold t_k. Then, by starting with a large threshold that decreases as the algorithm proceeds, we expect the algorithm to first look globally for good parts in the neighborhood graph and to gradually restrict its view so that it ultimately converges to a high-quality locally optimal solution. The resulting algorithm is called *threshold accepting*.

Simulated annealing is a randomized version of threshold accepting in which the threshold is a random variable. While in threshold accepting a fixed threshold t_k

is specified for each iteration k, we only specify the expected value c_k of t_k in simulated annealing. More specifically, the actual threshold used is taken randomly from the interval $(0, \infty)$ according to an exponential distribution with parameter $1/c_k$. By definition, this means that the distribution function of threshold t_k is given by $F(x) = \mathbb{P}\{t_i \leq x\} = 1 - \exp(-x/c_k)$ for $x > 0$. Using this distribution, it follows that if the cost of a solution s' is $\Delta = f(s') - f(s)$ larger than the cost of solution s, then the probability of accepting s' from s is given by

$$\mathbb{P}\{\text{accept } s'\} = \mathbb{P}\{\Delta < t_k\} = 1 - F(\Delta) = \exp\left(-\frac{\Delta}{c_k}\right)$$

if $\Delta > 0$ and by 1 if $\Delta \leq 0$. In other words, we obtain that a solution s' is accepted from s at iteration k with an *acceptance probability* that is given by

$$\mathbb{P}\{\text{accept } s'\} = \begin{cases} 1 & \text{if } f(s') \leq f(s), \\ \exp\left(\frac{f(s)-f(s')}{c_k}\right) & \text{if } f(s') > f(s). \end{cases}$$

The parameter c_k, which gives the expected value of the threshold, is called the *control parameter* or the *temperature*. The latter term is a result of the analogy with the physical annealing process discussed at the end of this section.

The difference in strategy for accepting non-improving moves in threshold accepting and simulated annealing is visualized in Figure 7.1. The former algorithm accepts a solution if and only if it yields an increase in cost of less than the threshold, while simulated annealing is capable of accepting any solution s', regardless of the resulting increase in cost. However, the larger the increase in cost, the smaller the probability that s' is indeed accepted by simulated annealing. To illustrate this we observe that, as the probability of accepting a cost increase of m times the expected value c_k of the threshold is given by e^{-m}, a deterioration, for instance, of $20c_k$ is accepted with a negligible frequency of about one in half a billion.

The simulated annealing algorithm is summarized in Figure 7.2. The algorithm starts by constructing an initial solution. From this solution a walk is initiated through the neighborhood graph. Being at solution s, it first selects a candidate solution s' from $N(s)$ to replace it. Generally, the candidate is taken uniformly at random from $N(s)$. If s' is better than s, then the algorithm moves to s'. Otherwise, it moves to s' with probability

$$\exp\left(\frac{f(s) - f(s')}{c_k}\right).$$

This can be implemented by comparing the value of this expression with a random number generated from the uniform distribution on the interval $[0, 1)$. The algorithm stops if a given stop criterion is met. A possible stop criterion used in practice is an upper bound on the number of successive iterations for the algorithm at the same solution value.

An important factor for the performance of simulated annealing is the *cooling schedule* used. The cooling schedule specifies the sequence c_1, c_2, \ldots of control parameter values. For threshold accepting, we have already argued that we prefer

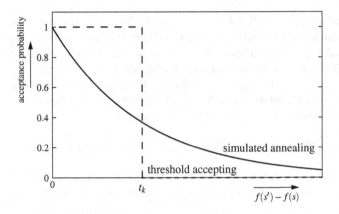

Figure 7.1. Acceptance probability of a solution s' from a solution s for simulated annealing and threshold accepting in the case where the expected value c_k of the randomized threshold used in simulated annealing is equal to the fixed threshold t_k used in threshold accepting.

to accept large deteriorations with a frequency that gradually decreases during the execution of the algorithm. Therefore, c_k is defined in such a way that it is non-increasing in k and that it approaches zero for $k \to \infty$. As simulated annealing reduces to iterative improvement for c_k close to zero, the algorithm ultimately ends up in a local optimum. We note that in many practical cooling schedules the control parameter is not decreased after each iteration, but is kept constant for a considerable number of iterations before it is decreased.

Unlike with iterative improvement, it is generally not useful to spend much time on constructing a high-quality initial solution as this solution has hardly any effect on the final solution. This follows from the observations that at high values of the control parameter the algorithm roughly resembles random search and that the diameter of a neighborhood graph is generally relatively small. In Theorems 4.2 and 4.4 the latter claim is substantiated for TSP with the node-insertion and 2-change neighborhood functions.

Simulated annealing finds its origin in the analogy between the solving of a combinatorial optimization problem and the physical annealing process of solids. In condensed matter physics, annealing is known as a thermal process for obtaining low energy states of a solid in a heat bath. The annealing process starts by melting the solid in the heat bath, in which case the particles are arranged randomly. Next, the temperature of the heat bath is carefully decreased, which results in a minimum energy state of the solid.

The evolution of a solid in a heat bath to thermal equilibrium is simulated by the *Metropolis algorithm*. We can use this algorithm for solving a combinatorial optimization problem by assuming the following equivalences between a combinatorial optimization problem and a physical many-particle system. The solutions of

algorithm Simulated annealing

begin
 $s :=$ some initial solution;
 $k := 1$;
 repeat
 generate an $s' \in N(s)$;
 if $f(s') \leq f(s)$ **then** $s := s'$
 else
 if $\exp\left(\frac{f(s)-f(s')}{c_k}\right) >$ random$[0,1)$ **then** $s := s'$;
 $k := k+1$;
 until stopcriterion;
end;

Figure 7.2. Simulated annealing algorithm for a minimization problem.

the optimization problem are given by the states of the system, and the cost of a solution is given by the energy of the corresponding state. This turns the Metropolis algorithm into an execution of simulated annealing in which the control parameter is assumed to be constant.

An appealing theoretical property of simulated annealing is that it can be proved to asymptotically converge to the set of globally optimal solutions. Proving this property is the subject of Chapter 8.

7.2 Tabu Search

Like simulated annealing, tabu search can in a sense be considered as a generalization of iterative improvement. However, the algorithms are based on a different pivoting rule. While simulated annealing uses a pivoting rule that resembles first improvement, tabu search applies a variant of the best-improvement pivoting rule. Hence, as a starting point, consider iterative improvement with the best-improvement pivoting rule.

An obvious strategy for the algorithm to escape from local optima is to drop the condition that a move is only performed if it leads to an improvement. This means that instead of stopping at a locally optimal solution, we move to a neighbor that results in the smallest increase in cost. However, a straightforward application of this strategy results in a high risk of returning to this same local optimum, either immediately or within a few steps. To restrict this undesirable behavior, which in Section 1.1 we called cycling, tabu search keeps a *tabu list T* of the recently visited solutions and it does not allow a return to any of the solutions in the list. In this way, cycles of length at most m can be prevented by using a tabu list of length m. We note that a trade-off has to be made when choosing m. Short tabu lists may not prevent cycling, whereas long tabu lists may lead to an excessive restriction of neighborhoods.

Although the tabu list approach discussed seems to be reasonable, it is generally not effective in practice since a lot of storage is required to keep a list of complete solutions and it is too time-consuming to check whether a solution is admissible in view of the high frequency at which the check has to be performed. One therefore often resorts to only keeping track of the last m transformations that have been performed and prohibiting the reverse of these transformations. For instance, for 2-change this means that the tabu list T contains the last m pairs of edges that have been selected for a 2-change. Alternatively, one can also store in T some key features of the last m solutions visited.

Obviously, the simplification of the tabu list results in a loss of information. It may no longer be guaranteed that a cycle of length at most m will not occur. A second negative consequence may be that unvisited solutions can obtain a tabu status. To prevent us from missing attractive solutions as a result of the latter problem, tabu search is equipped with a mechanism to overrule a tabu, called *aspiration criteria*. If a solution is tabu, then it may still be accepted if it satisfies one of the aspiration criteria that are defined for the algorithm. Almost all tabu search implementations use the obvious aspiration criterion that they may always move to a solution better than the best one found so far. Another useful aspiration criterion is that if all neighbors have a tabu status, which implies that we cannot move to a subsequent solution, then the tabu status of at least one neighbor is revoked.

We now arrive at the basic tabu search algorithm given in Figure 7.3, where a solution s' is said to be admissible if a move from s to s' is not prohibited by

algorithm Tabu search

begin
 $T := [\,]$;
 $s :=$ some initial solution;
 $s_{best} := s$;
 repeat
 find the best *admissible* $s' \in N(s)$;
 if $f(s') < f(s_{best})$ **then** $s_{best} := s'$;
 $s := s'$;
 update tabu list T;
 until stopcriterion;
end;

Figure 7.3. Tabu search algorithm for a minimization problem.

the tabu list T or if it is allowed because one of the aspiration criteria is satisfied. We emphasize that, unlike the iterative improvement and simulated annealing algorithms formalized in Figures 1.1 and 7.2, tabu search keeps track of the best solution encountered, s_{best}. For iterative improvement, this is obviously redundant. Furthermore, it is not essential for simulated annealing because the algorithm converges to the set of globally optimal solutions, which implies that it generally finds its best

solution at the end of its execution. For tabu search, however, it is an essential part of the algorithm. We nevertheless note that in simulated annealing implementations the best solution encountered is usually also stored as it is a potential improvement that can easily be implemented.

Simulated annealing may not be stopped at any desired moment in time. Because the control parameter has to converge to a value close to zero to obtain a meaningful implementation, the cooling schedule needs to be tuned to the time available for deriving a solution. Tabu search, on the other hand, can safely be interrupted, which means that it can use the amount of elapsed computing time as a stop criterion. An alternative and frequently used stop criterion for tabu search is an upper bound on the number of successive iterations in which no improvement is obtained over the best solution found.

A tabu list can be considered as a short-term memory of tabu search. Often, tabu search also makes use of a longer-term memory to achieve *intensification* and/or *diversification*. The aim of intensification is to explore more thoroughly the parts of the solution space that seem to contain promising solutions. For instance, if some attractive solutions derived by the algorithm show some common features, then this may give rise to the assumption that these features are characteristic of good solutions. To exploit this assumption, we can fix these features for a number of iterations to search more effectively for high-quality solutions.

The opposite of intensification is diversification. In spite of the tabu list, tabu search may suffer from the tendency to explore only a small part of the solution space, such that the part containing the really good solutions may be overlooked. Diversification aims to prevent this. Whether or not tabu search restricts itself to a particular part of the solution space is usually detected by recording the number of iterations that some features are present in the solution considered. We can enforce the algorithm to explore new areas of the solution space by, for instance, introducing rare features into the solution.

We note that we have already seen the basic idea of tabu search in Section 2.5, where the blocking rule plays the role of a tabu list in variable-depth search. We conclude our discussion on tabu search by emphasizing that, to make the approach successful, it is essential to devote serious effort to filling in the details of the algorithm in order to obtain a good performance.

7.3 Random Restart, GRASP, and Iterated Local Search

An obvious strategy for deriving higher-quality local optima than obtained by a single run of iterative improvement is to run the algorithm multiple times. We use the term *multistart* to denote algorithms that are based on this strategy. Examples of multistart algorithms are given in this and the next section.

A trivial extension of iterative improvement is *random restart*, in which multiple runs of iterative improvement are performed, each initialized with a different randomly selected starting solution. Although it is an adequate strategy for overcoming the problem of accidentally ending up in a poor local optimum, it does not help us to find one of the sparse local optima that are very close to optimality. For a single run

of iterative improvement we have already emphasized the importance of starting in a high-quality initial solution so that iterative improvement will explore a promising part of the solution space. This increases the chance of ending up in a high-quality local optimum, and it will reduce the running time of the algorithm. In conformity with this, we can improve random restart by using a more clever strategy for initializing the multiple runs of iterative improvement than just picking a random solution. We now discuss an example of multistart that is based on this idea.

Consider TSP and suppose that we only want to run iterative improvement once. Then, the nearest-neighbor algorithm is a candidate to construct an initial tour. This algorithm constructs a tour starting in city $c_1 = 1$. It extends the partially defined tour $\tau = (c_1)$ by repeatedly adding an unvisited city to the tour in the following way. In iteration i with $1 \leq i < n$ the tour $\tau = (c_1, c_2, \ldots, c_i)$ is extended with the city c_{i+1} not in τ that has the shortest distance from city c_i, i.e., for which $d_{c_i, c_{i+1}}$ is minimal.

Whether the nearest-neighbor algorithm is also applicable in multistart depends on the pivoting rule used in the iterative improvement implementation. If the pivoting rule used allows a random neighbor with better cost to be chosen, which is often the case when one applies first improvement, starting from the same solution can still result in different local optima. However, by initializing each run of iterative improvement with the same tour, we run the risk of exploring only a small part of the solution space. Therefore, it is often preferable to use a randomized algorithm that can generate different initial tours. Nearest neighbor can easily be transformed into such an algorithm by extending a partially defined tour $\tau = (c_1, c_2, \ldots, c_i)$ in iteration i with a random city from the candidate list $L_m(\tau)$ for some given positive integer m, where $L_m(\tau)$ contains the m cities not in τ that are closest to the last city c_i. By choosing m appropriately, diversity of the initial solutions can be traded off against their (expected) quality. In the two extreme cases, we find that a random tour is generated for $m = n$ and that the original deterministic algorithm is obtained for $m = 1$.

In the example discussed, we implement multistart by transforming a greedy constructive algorithm into a randomized algorithm by using candidate lists. In the literature, such a local search algorithm is also called *Greedy Randomized Adaptive Search Procedure (GRASP)*. Alternatively, we can also generate a new initial solution for iterative improvement by perturbing the locally optimal solution just derived. This is called *iterated local search*. For instance, when we have arrived at a 2-change optimal tour, we can perform an improving or random k-change with $k \geq 3$ to obtain a high-quality tour that we can use as the initial solution for the next run of iterative improvement.

As a generalization of iterated local search, we can also generate new initial solutions by combining multiple derived local optima. This is done in genetic local search, which is the subject of the next section.

7.4 Genetic Local Search

In this section we discuss genetic local search, which is a genetic algorithm incorporating local search. As a genetic algorithm is inspired by the natural evolution

process, we first discuss briefly how it is generally believed that species evolve over time. The two key mechanisms in evolution are *selection* and *variation*. Selection is achieved in nature by a process called survival of the fittest, meaning that the individuals of a population that are better adjusted to their habitat have a higher chance of surviving. These individuals also have a higher chance of mating, which implies that their genotype is more likely to remain in successive generations of the population, where the genotype of individuals refers to the inheritable information stored in chromosomes. As a result, the population moves towards a better overall fitness.

Variation causes new, hopefully better, genotypes to be added to the population. Variation is established by means of *recombination* and *mutation*. Recombination is a result of sexual reproduction of two parents and corresponds to combining the genotype of the two parents into a new genotype for their child. Mutation, on the other hand, is asexual and corresponds to random perturbations of the genotype of an individual. Hence, while recombination mixes the genetic material that occurs in a population, mutation can create new genetic material.

Genetic algorithms are based on natural evolution, where each individual represents a solution and the fitness of an individual is given by the cost of the corresponding solution. The algorithm starts by generating N initial solutions for a given positive N. This set forms the initial population P_0. The genetic algorithm now repeatedly generates a new population of size N until a given stop criterion is met. Let P_t be the population derived in iteration t. Population P_t is obtained from P_{t-1} by first selecting a subset of the solutions in P_{t-1}. This selection is guided by the cost function. Often it is done probabilistically, which means that the better a solution the higher its chance of being selected. Next, new solutions are created from the selected ones by recombination and/or mutation. As recombination operator we can use any algorithm that constructs a solution from a number of parent solutions, where the number of parents may be larger than two, and the mutation operator corresponds to an algorithm for changing a single solution. The new solutions are added to the population. If population P_t has size N again, then its construction has been completed. As the stopping criterion for the algorithm we can use an upper bound on the number of populations generated. Alternatively, one can, for instance, assume a maximum number of iterations in which no improvement on the best solutions is found.

In the algorithm described, offspring solutions obtained by recombination and/or mutation are added to a population without removing their parent solutions. It is also possible that the set of offspring solutions substitutes some or all parent solutions from which they originate. As a solution may be selected for recombination multiple times, this need not conflict with keeping the population size constant.

We can use genetic algorithms to derive high-quality local optima if we define the mutation operator as an execution of iterative improvement and the recombination operator as an algorithm for combining two or more local optima into a new solution. The resulting algorithm, which is called *genetic local search algorithm* or *memetic algorithm*, can be considered as multistart, where except for the first N runs, each run of iterative improvement is initialized with a solution that is constructed

from previously derived locally optimal solutions by means of the recombination algorithm. By choosing a recombination algorithm that produces offspring solutions that are genetically closely related to their parents, we expect that the cost of an offspring is not too much higher than the cost of its parents. Hence, in that case all but the first N runs of iterative improvements are initialized with good quality solutions. The genetic local search algorithm is given in Figure 7.4. Note that in the algorithm an offspring solution is added to the population if it is obtained by recombination and that it substitutes its parent solution if it is obtained by mutation. In the

algorithm Genetic local search

begin
 $P_0 :=$ set of N solutions;
 /* *Mutation* */
 replace each $s \in P_0$ by Iterative_Improvement(s);
 $t := 1$;
 repeat
 Select $P_t \subseteq P_{t-1}$;
 /* *Recombination* */
 extend P_t by adding offspring;
 /* *Mutation* */
 replace each $s \in P_t$ by Iterative_Improvement(s);
 $t := t + 1$;
 until stopcriterion;
end;

Figure 7.4. Genetic local search algorithm for a minimization problem.

following two examples we give examples of possible recombination operators for TSP.

Example 7.1. We can combine two parent tours τ_1 and τ_2 into a single offspring tour τ in the following way, where we assume without loss of generality that city 1 is the first city in both tours. First of all, a random subpath p in τ_2 is selected. The tour τ is then obtained by first traversing path p and next visiting all other cities in the order in which they appear in τ_1. The first city that is visited after p is chosen randomly. If we consider SYMMETRIC TSP, then we choose the direction of path p and tour τ_1 at random. Figure 7.5 visualizes this recombination for tours $\tau_1 = (1,2,3,4,5,6,7,8)$ and $\tau_2 = (1,5,8,6,2,4,7,3)$ and path $p = (4,7,3)$. □

Example 7.2. In Section 7.2 we have already mentioned that if good solutions share some common features, then this indicates that these features might be characteristic of good solutions. Based on this observation, we can design a recombination algorithm that generates an offspring tour τ from any $m \geq 2$ parent tours $\tau_1, \tau_2, \ldots, \tau_m$ as follows. Let τ_1 have the shortest length among all m parent tours. Then, we let tour τ contain an arc a from τ_1 with a probability p_a that depends on whether the other

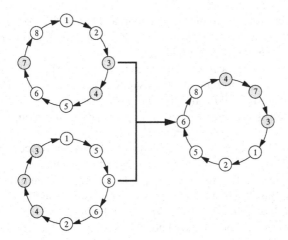

Figure 7.5. Example of recombining two tours into one new tour.

$m - 1$ parent tours have the arc. More precisely, if a tour τ_i with $i > 1$ also contains arc a, then this results in an increase in p_a by an amount that depends on the quality of tour τ_i, i.e., the shorter tour τ_i, the larger the effect of τ_i on p_a. This results in a partial tour consisting of several tour segments. These segments are then connected randomly to construct a complete tour. □

7.5 Bibliographical Notes

As mentioned in Section 7.1, simulated annealing is inspired by the Metropolis algorithm, which simulates the evolution of a solid in a heat bath at a given temperature. The Metropolis algorithm dates back to 1953, when it was introduced by Metropolis, Rosenbluth, Rosenbluth, Teller & Teller [1953]. Kirkpatrick, Gelatt & Vecchi [1983] and, independently, Cerny [1985] were the first to use simulated annealing in combinatorial optimization. Although we presented simulated annealing as an elaboration of threshold accepting, this is not in conformity with its introduction. Actually, Dueck & Scheuer [1990] presented threshold accepting as a simplification of simulated annealing. For a more thorough discussion on simulated annealing we refer you to Aarts & Korst [1989] and Salamon, Sibani & Frost [2002], while for threshold accepting we refer you to Winker [2001].

As a second metaheuristic, we proposed tabu search. Although many of its elements were already proposed before, the paper by Glover [1986] was the first to mention tabu search explicitly. Independently, Hansen [1986] proposed a similar approach, which he called steepest ascent mildest descent. The book by Glover & Laguna [1997] is devoted entirely to tabu search.

The paper of Lin & Kernighan [1973], to which we referred in the bibliographical notes of Chapter 2, is an early paper that proposes random restart. The authors use it in combination with a variable-depth search approach for tackling TSP. As another example of multistart, we mentioned GRASP in Section 7.3. Feo & Re-

sende [1989] were the first to define this approach explicitly. For more details on GRASP we refer to a chapter of Pitsoulis & Resende [2002] and to the annotated bibliography given by Festa & Resende [2002].

A survey on iterative local search is given by Lorenco, Martin & Stützle [2003]. The most well-known representative of this approach is the iterated Lin-Kernighan algorithm by Johnson [1990]. In this algorithm a new run of the Lin-Kernighan algorithm for TSP is obtained by performing a random improving double-bridge move, which is a special type of 4-change.

Mühlenbein, Gorges-Schleuter & Krämer [1988] introduced genetic local search. They applied it to TSP, where they used the recombination operator presented in Example 7.1. This recombination operator has been empirically analyzed by Ulder, Aarts, Bandelt, Van Laarhoven & Pesch [1991]. The recombination operator proposed in Example 7.2 is according to Schilham [2001].

Genetic algorithms were already known before genetic local search. The book by Holland [1975] is considered an important early contribution to the field. For extensive treatments on genetic algorithms we refer you to Goldberg [1989] and Michalewicz [1992], while genetic local search is discussed in more detail in Corne, Dorigo & Glover [1999].

Osman & Laporte [1996] present a metaheuristic bibliography with a classification containing more than 1,400 references both on the theory and application of metaheuristics. More recent overviews of metaheuristics are presented by Ribeiro & Hansen [2002] and Glover & Kochenberger [2003]. We conclude with referring to the books by Aarts & Lenstra [1997] and Michalewicz & Fogel [2000], which each give a clear and more elaborate presentation of simulated annealing, tabu search, genetic algorithms, and other metaheuristics.

7.6 Exercises

1. Consider the variant of threshold accepting in which we keep the threshold at a constant value t.

 a) Give the weakest condition on t under which this variant of threshold accepting corresponds to random search.

 b) Give the weakest condition on t under which this variant of threshold accepting corresponds to iterative improvement.

2. Give at least one useful aspiration criterion that is different from the ones given in Section 7.2.

3. In Section 2.5 we applied variable-depth search to UGP. The resulting neighborhood function was called Kernighan-Lin. Describe iterative improvement with this Kernighan-Lin neighborhood function in terms of tabu search.

4. Consider tabu search with a tabu list that keeps the m most recently visited solutions. Show that for any size of the solution space there exists a neighborhood graph for which tabu search converges to the worst $m + 2$ solutions.

5. Consider an iterated local search implementation for TSP that applies a single random 4-change whenever iterative improvement using the 2-change neighborhood function reaches a local optimum. Compare this approach with (a) iterative improvement using 2-change and (b) iterative improvement using 4-change. Discuss the advantages and disadvantages of the iterated local search approach compared to these iterative improvement implementations.

8

Asymptotic Convergence of Simulated Annealing

In Chapter 7 we presented several popular metaheuristics for finding high-quality local optima. Their popularity can be attributed mainly to the many practical successes they achieved for a wide variety of problems. An interesting question is whether the good performance of the metaheuristics can also be supported by theoretical results. This is the case for simulated annealing, for which it has been shown that it asymptotically converges to the set of globally optimal solutions if it is applied on a neighborhood graph that is finite, symmetric, and strongly connected. This chapter is devoted to proving this result. In Section 8.6 we show that the result does not necessarily hold for tabu search.

To put the convergence result of simulated annealing in the right perspective, we stress that it only states that the algorithm reaches a global optimum with probability 1 after an infinite number of iterations. This means that the simple total enumeration algorithm, which takes exponential but finite time, is more efficient when optimality has to be guaranteed. Furthermore, if the naive random search algorithm discussed in Section 7.1 stores its best solution encountered, then it also converges to a globally optimal solution. The relevance of the convergence result for simulated annealing is more subtle, however. It states that, when moving through a neighborhood graph, simulated annealing searches its way to the set of globally optimal solutions. This supports the assumption that the applied search strategy is effective.

In this chapter we denote the problem instance on which simulated annealing is applied by (S, f) and the neighborhood function used by N. Although the definition of a combinatorial optimization problem allows S to be infinite, we assume S to be finite in this chapter. Note that S is finite for all problems in NPO. Furthermore, we assume without loss of generality that we are dealing with a minimization problem and that $S = \{1, 2, \ldots, |S|\}$.

8.1 Mathematical Modeling

To prove the asymptotic convergence of simulated annealing (when applied to S, f, and N), we use the fact that a run of the algorithm can be modeled as a Markov chain. Below we give some relevant definitions from Markov theory.

Definition 8.1. A *discrete-time stochastic process* is a set of random variables $\{X(t) \mid t \in T\}$, where T is a finite or countably infinite set of time points. If $X(t) = i$, then we say that the process is in state i at time t. The set of all possible states is called the *state space*. □

Definition 8.2. A *Markov chain* is a discrete-time stochastic process $\{X(t) \mid t \in T\}$ with $T = \{0, 1, \ldots\}$ either finite or infinite, and it has a finite or countably infinite state space. Random variable $X(0)$ is independent of any other random variable, and $X(t)$ with $t \geq 1$ only depends on $X(t-1)$. Hence, if we are at time $t-1$, the next state $X(t)$ only depends on the current state $X(t-1) = i$ and not on how we arrived in state i. Without loss of generality, we assume that the state space is given by $\{1, 2, \ldots\}$. For any $t \geq 1$ and any pair i, j of states, the *transition probability* $P_{ij}(t)$ is defined as the probability that, if we are at time $t-1$, the next state is j given that the current state is i, i.e., $P_{ij}(t) = \mathbb{P}\{X(t) = j \mid X(t-1) = i\}$. The matrix $P(t)$, in which the (i, j)th element equals the transition probability $P_{ij}(t)$ is called the *transition matrix*. □

Definition 8.3. A Markov chain is said to be *finite* if it is defined on a finite state space. □

Definition 8.4. If the transition probabilities in a Markov chain are independent of the time point t, i.e., if $P(t) = P(t')$ for all $t, t' \geq 0$, then the Markov chain is said to be *homogeneous*. Otherwise, it is *inhomogeneous*. □

A convenient representation of a transition matrix P is given by the labeled directed graph $D = (V, A)$, in which the node set V corresponds to the state space of the Markov chain and which contains a directed arc (i, j) if and only if $P_{ij} > 0$. The label of an arc (i, j) is defined as P_{ij}. We call D the *state transition graph* of P. To illustrate this, Figure 8.1 gives the state transition graph defining the transition matrix

$$P = \begin{pmatrix} \frac{1}{4} & \frac{3}{4} & 0 \\ \frac{1}{3} & 0 & \frac{2}{3} \\ 0 & 0 & 1 \end{pmatrix}. \tag{8.1}$$

Figure 8.1. State transition graph corresponding to transition matrix 8.1.

Observation. A homogeneous Markov chain models a random walk in the state transition graph associated with its transition matrix where, at node i, an arc (i, j) is taken with a probability given by its label. □

Let $a_i(t)$ with $t \geq 0$ define the probability of being in state i at time t for a given Markov chain, i.e., $a_i(t) = \mathbb{P}\{X(t) = i\}$. Then we have

$$a_i(t) = \sum_{l \in \{1,2,\ldots\}} a_l(t - 1)P_{li}(t),$$

where $\{1, 2, \ldots\}$ denotes the state space. Hence, the vector $a(t) = (a_1(t), a_2(t), \ldots)$ that specifies the probability distribution of the states at time t, can be derived from the initial probability distribution $a(0)$ via

$$a(t) = a(0) \prod_{l=1}^{t} P(l). \tag{8.2}$$

For example, if in the homogeneous Markov chain based on the state transition graph of Figure 8.1 the initial state is selected uniformly at random, which means $a(0) = (\frac{1}{3}, \frac{1}{3}, \frac{1}{3})$, then the probability distribution after five time units is given by $a(0)P^5 \approx (0.03, 0.04, 0.93)$. Hence, if we perform a random walk of five steps in the transition graph of Figure 8.1, then we have a probability of 0.93 of ending up in state three.

The simulated annealing algorithm continuously attempts to replace a current solution by one of its neighbors. One such attempt is called a *transition*. This process can be viewed as a finite inhomogeneous Markov chain. The state space corresponds to the set S of solutions, random variable $X(0)$ gives the starting solution, random variable $X(k)$ with $k \geq 1$ gives the solution visited in the kth iteration of simulated annealing, and a single step in the Markov chain corresponds to a transition of the simulated annealing algorithm. To use the concept of a Markov chain, we need a transition in simulated annealing to depend only on the current solution, which is indeed the case.

The transition matrix associated with a run of simulated annealing is determined by the strategy used for generating neighboring solutions and for accepting a solution once it is generated. These two strategies are specified by the generation and acceptance matrices, respectively. The transition and acceptance matrices depend on the iteration number k via the control parameter c_k and the generation matrix is independent of k. This is made explicit in our notation.

Definition 8.5. The *generation matrix* G defines for each pair of solutions $i, j \in S$ the probability of generating solution j from solution i. An entry G_{ij} is called the *generation probability* and satisfies $G_{ij} > 0$ if and only if $j \in N(i)$. □

Definition 8.6. The *acceptance matrix* $A(c_k)$ defines for each pair of solutions $i, j \in S$ the probability of accepting solution j from solution i in the kth iteration of the simulated annealing algorithm. Specifically, we have that the *acceptance probability* $A_{ij}(c_k)$ is given by

$$A_{ij}(c_k) = \begin{cases} 1 & \text{if } f(j) \le f(i), \\ \exp\left(\frac{f(i)-f(j)}{c_k}\right) & \text{if } f(j) > f(i). \end{cases} \tag{8.3}$$

□

Definition 8.7. The *transition matrix* $P(c_k)$ defines for each pair of solutions $i, j \in S$ the probability of moving from solution i to solution j in the kth iteration of the simulated annealing algorithm. The *transition probability* $P_{ij}(c_k)$ is given by

$$P_{ij}(c_k) = \begin{cases} G_{ij}A_{ij}(c_k) & \text{if } i \ne j, \\ 1 - \sum_{l \in S, l \ne i} P_{il}(c_k) & \text{if } i = j. \end{cases}$$

□

We note that the generation and transition matrices are stochastic, while the acceptance matrix is not stochastic.

Definition 8.8. A vector $v = (v_1, v_2, \ldots, v_n)$ is called *stochastic* if its components are non-negative and add up to one, i.e., $v_i \ge 0$ for all i and $\sum_{i=1}^{n} v_i = 1$. A matrix is called stochastic if each row is a stochastic vector. □

Example 8.1. To illustrate the modeling of simulated annealing as a Markov chain consider the application of the algorithm on the neighborhood graph given in Figure 8.2(a). An obvious strategy for generating neighboring solutions is to generate all neighbors of a solution i with an equal probability of $1/|N(i)|$. This results in the generation matrix

$$G = \begin{pmatrix} 0 & 0 & \frac{1}{2} & \frac{1}{2} \\ 0 & 0 & \frac{1}{2} & \frac{1}{2} \\ \frac{1}{2} & \frac{1}{2} & 0 & 0 \\ \frac{1}{2} & \frac{1}{2} & 0 & 0 \end{pmatrix},$$

where the four solutions are numbered such that $f(i) = i$. The generation matrix is visualized in Figure 8.2(b). The acceptance matrix

$$A(c) = \begin{pmatrix} 1 & \exp\left(-\frac{1}{c}\right) & \exp\left(-\frac{2}{c}\right) & \exp\left(-\frac{3}{c}\right) \\ 1 & 1 & \exp\left(-\frac{1}{c}\right) & \exp\left(-\frac{2}{c}\right) \\ 1 & 1 & 1 & \exp\left(-\frac{1}{c}\right) \\ 1 & 1 & 1 & 1 \end{pmatrix}$$

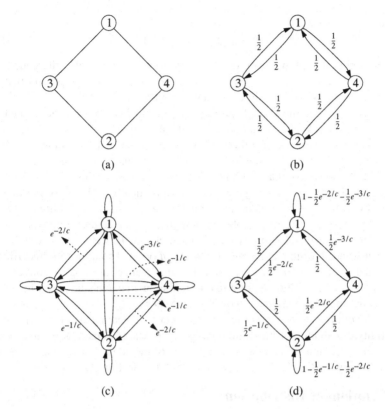

Figure 8.2. (a) Neighborhood graph, where the numbers inside the nodes indicate the cost of the solutions. (b) Generation matrix. (c) Acceptance matrix, where the label of an arc is only given if it is different from one. (d) Transition matrix.

follows directly from (8.3) and is depicted in Figure 8.2(c). Using G and $A(c)$, we can now determine the transition matrix

$$P(c) = \begin{pmatrix} \alpha_1(c) & 0 & \frac{1}{2}\exp\left(-\frac{2}{c}\right) & \frac{1}{2}\exp\left(-\frac{3}{c}\right) \\ 0 & \alpha_2(c) & \frac{1}{2}\exp\left(-\frac{1}{c}\right) & \frac{1}{2}\exp\left(-\frac{2}{c}\right) \\ \frac{1}{2} & \frac{1}{2} & 0 & 0 \\ \frac{1}{2} & \frac{1}{2} & 0 & 0 \end{pmatrix},$$

where $\alpha_1(c)$ is given by $1 - \frac{1}{2}\exp\left(-\frac{2}{c}\right) - \frac{1}{2}\exp\left(-\frac{3}{c}\right)$ and $\alpha_2(c)$ is given by $1 - \frac{1}{2}\exp\left(-\frac{1}{c}\right) - \frac{1}{2}\exp\left(-\frac{2}{c}\right)$. The associated state transition graph is given in Figure 8.2(d). At a fixed value of the control parameter, a run of simulated annealing now corresponds to a random walk in this state transition graph. □

This concludes the mathematical modeling of simulated annealing. We can now formalize the asymptotic convergence of simulated annealing to the set $S^* \subseteq S$ of

optimal solutions by

$$\lim_{k \to \infty} \mathbb{P}\{X(k) \in S^*\} = 1. \tag{8.4}$$

It is our goal in this chapter to prove this equation for the case that the generation matrix implies a symmetric, strongly connected neighborhood graph and that the control parameter is lowered slowly enough with $\lim_{k \to \infty} c_k = 0$.

In Section 7.1 we mentioned that in many practical cooling schedules the control parameter c_k is kept constant for a considerable number of iterations after each decrease in value. In this case the algorithm can be viewed as a sequence of homogeneous Markov chains, one for each different value of the control parameter. In Section 8.3 we prove that if the homogeneous Markov chains are all of infinite length, then simulated annealing converges asymptotically to the set of optimal solutions. In Section 8.4 we then derive some statistical results for this homogeneous model. Of course, it is not possible to implement simulated annealing such that it consists of a sequence of infinitely long Markov chains. However, in Section 8.5 we show that if the algorithm converges to the set of optimal solutions in this artificial model, then we can define a cooling schedule that realizes asymptotic convergence to the set of optimal solutions for a real run of simulated annealing, i.e., for which the resulting inhomogeneous Markov chain satisfies (8.4).

The proof of the convergence result of simulated annealing uses the fact that the probability distribution of the finite homogeneous Markov chain associated with a run of simulated annealing at some fixed value of the control parameter converges to a 'stationary distribution'. This claim is proved in Section 8.2.

8.2 Stationary Distribution

Consider the state transition graph of Figure 8.1. Starting in state three, the corresponding Markov chain always remains in this state. This means that the probability distribution $q = (0, 0, 1)$ describes a stable situation: once a random variable $X(t)$ in the Markov chain satisfies this distribution, all subsequent random variables $X(t')$ with $t' \geq t$ satisfy this distribution as well. Such a distribution is said to be stationary.

Definition 8.9. A stochastic vector q is a *stationary distribution* of a homogeneous Markov chain with transition matrix P in the case that the following property holds. If $X(0)$ has probability distribution q, then each random variable $X(t)$ with $t \geq 0$ has probability distribution q. Formally, this means that q satisfies $q = qP$, i.e., q is the left eigenvector of P with eigenvalue 1. □

Theorem 8.1. *A finite homogeneous Markov chain has at least one stationary distribution.*

Proof. See, for instance, Theorem 8.3.11 in Berman & Plemmons [1979]. □

Using (8.2) it can easily be verified that if the probability distribution of a finite homogeneous Markov chain converges to a given distribution q, i.e., if $q = \lim_{t \to \infty} a(t)$

exists, then q has to be a stationary distribution:

$$q = \lim_{t\to\infty} a(t) = \lim_{t\to\infty} a(0)P^t = \lim_{t\to\infty} a(0)P^{t-1}P$$
$$= \lim_{t'\to\infty} a(0)P^{t'}P = qP.$$

Suppose that we apply simulated annealing to a strongly connected neighborhood graph. We show for the probability distribution of the homogeneous Markov chain associated with the algorithm at a given value of the control parameter that it not only converges to a stationary distribution, but that this stationary distribution is also independent of the initial probability distribution $a(0)$ of the states (solutions). More formally, we show that the Markov chain is 'strongly ergodic', which means the following.

Definition 8.10. A Markov chain is called *strongly ergodic* if a stochastic vector q exists, such that for all $t \geq 0$ and each pair i, j of states we have

$$q_i = \lim_{t'\to\infty} \mathbb{P}\{X(t') = i \mid X(t) = j\}. \tag{8.5}$$

\square

As mentioned above, a finite homogeneous Markov chain has the property that if it is strongly ergodic, then the probability distribution to which it converges is a stationary one. The following theorem states that this stationary distribution is also the only stationary distribution of the Markov chain.

Theorem 8.2. *If a finite homogeneous Markov chain is strongly ergodic, then the stochastic vector q to which its probability distribution converges is its only stationary distribution.*

Proof. Above we showed that if the probability distribution of a homogeneous Markov chain converges to a distribution, then this distribution has to be a stationary one. Hence, to prove the theorem, it remains to be shown that an arbitrary, strongly ergodic, finite, homogeneous Markov chain does not have multiple stationary distributions. This result follows from the observations that the probability distribution of a strongly ergodic Markov chain converges to a unique distribution q and that for a stationary distribution q' it holds that if the initial probability distribution of the Markov chain is given by q', then the probability distribution of the Markov chain remains and thus converges to this distribution, i.e., $q = q'$. \square

In the definition of strong ergodicity, we require that the convergence to a given distribution q is guaranteed not only for any initial probability distribution $a(0)$, but also for any intermediate probability distribution $a(t)$ with $t \geq 1$. For homogeneous Markov chains this is not a real additional constraint as their behavior does not depend on the time unit at which they start. The constraint does, however, make sense for inhomogeneous Markov chains, which will be discussed later on.

To prove the existence of a q satisfying (8.5) for a given homogeneous Markov chain, the following two definitions are useful.

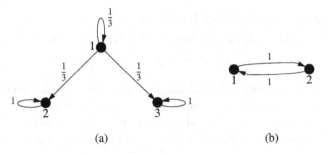

(a) (b)

Figure 8.3. (a) State transition graph of a decomposable Markov chain that does not have a unique stationary distribution to which its probability distribution converges. (b) State transition graph of a non-convergent periodic Markov chain.

Definition 8.11. A homogeneous Markov chain with transition matrix P is called *irreducible* if, for each pair i, j of states, the state transition graph contains a path from i to j, which is equivalent to saying that an $n \geq 1$ exists with $(P^n)_{ij} > 0$. □

Definition 8.12. A homogeneous Markov chain with transition matrix P is called *aperiodic* if for each state i the greatest common divisor $\gcd(W_i) = 1$, where W_i is the set containing the lengths of all paths from i to itself in the state transition graph, i.e., $n \in W_i$ if and only if $(P^n)_{ii} > 0$. □

As an example, observe that the homogeneous Markov chain based on the state transition graph given in Figure 8.1 is not irreducible as the transition graph contains no path from state three to any other state. It is, however, aperiodic as $W_1 = W_3 = \mathbb{N}^+$ and $W_2 = \mathbb{N}^+ \setminus \{1\}$.

To see that the above two definitions are relevant to ensure convergence to a unique stationary distribution, consider the two state transition graphs given in Figure 8.3. The state transition graph of Figure 8.3(a) defines an aperiodic Markov chain that is decomposable, which means that it is not irreducible. Once the Markov chain is in state two or three, it always stays in that state. From this observation it follows that the probability distribution of the Markov chain converges to the stationary distribution $(0, \frac{1}{2}, \frac{1}{2})$ if we start in state one, to stationary distribution $(0, 1, 0)$ if we start in state two, and to stationary distribution $(0, 0, 1)$ if we start in state three.

Next, consider the irreducible Markov chain defined by the state transition graph of Figure 8.3(b). This Markov chain is periodic as all paths in the graph from a node to itself have even length. Unlike the Markov chain associated with the state transition graph of Figure 8.3(a), this Markov chain does have a unique stationary distribution, namely $q = (\frac{1}{2}, \frac{1}{2})$. However, it converges to q only if q is also the initial distribution of the states, i.e., if $a(0) = q$ holds. For any other initial distribution $a(0) = (p, 1 - p)$ the distribution of the states alternates between $(p, 1 - p)$ and $(1 - p, p)$.

The following theorem is proved in many books on Markov theory and states that irreducibility and aperiodicity are sufficient conditions for a finite homogeneous

Markov chain to guarantee convergence to a unique stationary distribution.

Theorem 8.3. *If a finite homogeneous Markov chain is both irreducible and aperiodic, then it is strongly ergodic.*

Proof. For proof, we refer to Feller [1950], Isaacson & Madsen [1976], and Grimmett & Stirzaker [1992]. $\qquad\square$

From the theorem it follows that to prove the asymptotic convergence of simulated annealing to a unique stationary distribution in the case that the control parameter is assumed to be fixed, it suffices to show that the associated Markov chain is both irreducible and aperiodic. To prove this, the following lemma is useful. It states that once a Markov chain has been proved to be irreducible, we only have to check whether the state transition graph contains at least one loop to conclude that it is also aperiodic.

Lemma 8.1. *An irreducible finite homogeneous Markov chain with transition matrix P is aperiodic if a state i exists with $P_{ii} > 0$.*

Proof. Consider an arbitrary irreducible Markov chain with $P_{ii} > 0$ for a state i. To prove the lemma we have to show that $\gcd(W_j) = 1$ for any state j. As the Markov chain is irreducible, we have that $k, l \in \mathbb{N}^+$ exist such that $P_{ji}^k > 0$ and $P_{ij}^l > 0$. Hence, we get $P_{jj}^{k+l} \geq P_{ji}^k P_{ij}^l > 0$. Furthermore, since $P_{ii} > 0$ we also have $P_{jj}^{k+l+1} \geq P_{ji}^k P_{ii} P_{ij}^l > 0$. This yields that $k+l, k+l+1 \in W_j$, which implies that indeed $\gcd(W_j) = 1$. $\qquad\square$

Theorem 8.4. *Let (S, f) be an instance of a combinatorial optimization problem with $S^* \neq S$ and S finite. Furthermore, let N be a neighborhood function defined on S that induces a strongly connected (but not necessarily symmetric) neighborhood graph.*

The finite homogeneous Markov chain associated with a run of simulated annealing at a fixed value c of the control parameter is strongly ergodic, i.e., a unique stationary distribution $q(c)$ exists to which the probability distribution of the Markov chain converges.

Proof. To prove the theorem it suffices by Theorem 8.3 to show that the Markov chain is both irreducible and aperiodic. We start by proving the former property. Let $i, j \in S$ be arbitrary. We have to prove $P_{ij}^n(c) > 0$ for an $n \geq 1$. As the neighborhood graph is strongly connected, Definition 8.5 gives that a sequence $s_1, s_2, \ldots, s_n \in S$ of solutions exists, such that $s_1 = i$, $s_n = j$, and $G_{s_l, s_{l+1}} > 0$ for all $1 \leq l < n$. Furthermore, by Definition 8.6 we have $A_{s,s'}(c) > 0$ for any two solutions s, s'. As a result, we obtain

$$P_{ij}^n(c) \geq \prod_{l=1}^{n-1} G_{s_l, s_{l+1}} A_{s_l, s_{l+1}}(c) > 0.$$

Thus, the Markov chain is indeed irreducible.

To prove that the Markov chain is also aperiodic, it suffices by Lemma 8.1 to show that a solution i exists with $P_{ii}(c) > 0$. Since the neighborhood graph is strongly connected and since $S^* \neq S$, solutions $i, j \in S$ exist with $f(i) < f(j)$ and

$G_{ij} > 0$, i.e., $j \in N(i)$. By the definition of P, we have

$$P_{ii}(c) = 1 - \sum_{l \in S, l \neq i} G_{il} A_{il}(c).$$

As $A_{il}(c) \leq 1$ for all $l \in S$, this gives

$$P_{ii}(c) \geq 1 - G_{ij} A_{ij}(c) - \sum_{l \in S, l \neq i, j} G_{il}.$$

From $f(i) < f(j)$ it follows that $A_{ij}(c) < 1$. Hence, as by assumption $G_{ij} > 0$, we obtain

$$P_{ii}(c) > 1 - \sum_{l \in S, l \neq i} G_{il} = 0.$$

□

In order for Theorem 8.4 to be applied, the neighborhood graph needs to be strongly connected. Because the theorem serves as an auxiliary result for proving convergence of simulated annealing to the set of optimal solutions, it is interesting to consider the question of whether the condition is also relevant in the context of this convergence result. The answer to this question is in the affirmative because if an optimal solution is unreachable from a solution, then convergence to the set of optimal solutions can obviously not be achieved.

8.3 Homogeneous Markov Chains

Theorem 8.4 states that the probability distribution $a(t)$ of the solution space S converges to a unique stationary distribution $q(c)$ if simulated annealing is run at a fixed value c of the control parameter. In this section we prove that simulated annealing converges to the set of optimal solutions in the case that the value of the control parameter c approaches zero and is only changed after equilibrium is attained, i.e., after the algorithm has converged to the stationary distribution $q(c)$. Formally, this means that we prove

$$\lim_{c \downarrow 0} q_i(c) = 0 \quad \text{for all } i \in S \setminus S^*. \tag{8.6}$$

This model of simulated annealing is not a realistic one as equilibrium is only attained after an infinite number of iterations of simulated annealing. In other words, the model views simulated annealing as a sequence of homogeneous Markov chains of infinite length. However, the result already provides some evidence that the algorithm will also converge in a realistic setting. In Section 8.5 this claim is substantiated by proving that convergence in the homogeneous model implies the existence of a cooling schedule that realizes convergence for a real run of simulated annealing.

At the end of Section 8.2 we indicated the necessity of assuming the neighborhood graph to be strongly connected in order to have an asymptotic convergence of simulated annealing to the set of globally optimal solutions. This condition alone is not sufficient to guarantee convergence in the homogeneous model, however. This follows from the following example.

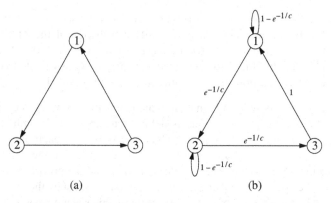

Figure 8.4. (a) Neighborhood graph, where the numbers inside the nodes indicate the cost of the solutions. (b) State transition graph of the transition matrix following from the application of simulated annealing on this neighborhood graph.

Example 8.2. Consider the strongly connected neighborhood graph given in Figure 8.4(a). The solutions are numbered 1,2, and 3, and the cost of a solution is given by its number. When simulated annealing is applied to this neighborhood graph, the generation matrix is given by

$$G = \begin{pmatrix} 0 & 1 & 0 \\ 0 & 0 & 1 \\ 1 & 0 & 0 \end{pmatrix},$$

the acceptance matrix by

$$A(c) = \begin{pmatrix} 1 & \exp\left(-\frac{1}{c}\right) & \exp\left(-\frac{2}{c}\right) \\ 1 & 1 & \exp\left(-\frac{1}{c}\right) \\ 1 & 1 & 1 \end{pmatrix},$$

and the transition matrix by

$$P(c) = \begin{pmatrix} 1 - \exp\left(-\frac{1}{c}\right) & \exp\left(-\frac{1}{c}\right) & 0 \\ 0 & 1 - \exp\left(-\frac{1}{c}\right) & \exp\left(-\frac{1}{c}\right) \\ 1 & 0 & 0 \end{pmatrix}.$$

The state transition graph corresponding to $P(c)$ is depicted in Figure 8.4(b). By Theorem 8.4, the homogeneous Markov chain associated with $P(c)$ has a unique stationary distribution to which its probability distribution converges, i.e., the Markov chain converges in distribution to the only existing stochastic left eigenvector q of $P(c)$ with eigenvalue 1. It can be verified that

$$q(c) = \left(\frac{1}{2+\exp\left(-\frac{1}{c}\right)}, \frac{1}{2+\exp\left(-\frac{1}{c}\right)}, \frac{\exp\left(-\frac{1}{c}\right)}{2+\exp\left(-\frac{1}{c}\right)} \right)$$

is a left stochastic eigenvector of $P(c)$ with eigenvalue 1 and thus also the unique stationary distribution to which the probability distribution of the Markov chain converges. For $c \downarrow 0$, stationary distribution $q(c)$ approaches $(\frac{1}{2}, \frac{1}{2}, 0)$. Hence, (8.6) does not hold: in the limiting case simulated annealing is equally likely to end up in solution 2, which is suboptimal, as in solution 1, which is optimal. □

The main result of this section is that simulated annealing converges to the set of optimal solutions in the homogeneous model if the neighborhood graph is not only strongly connected but also symmetric, which means that for each pair $i, j \in S$ of solutions we have $G_{ij} > 0$ if and only if $G_{ji} > 0$. Observe that the neighborhood graph of Figure 8.4 is asymmetric. To better understand the proof, we first prove the result for the special case that the generation matrix satisfies the stronger condition $G_{ij} = G_{ji}$ for all $i, j \in S$. As an example of this special case, consider the 2-change neighborhood function for TSP. As indicated in Section 2.1, each tour has the same neighborhood size of $2 + n(n - 3)$ with respect to this neighborhood function. Hence, if a neighbor is selected uniformly at random, then $G_{ij} = G_{ji}$ holds for all $i, j \in S$.

The following theorem gives us an important auxiliary result.

Theorem 8.5. *Let (S, f) be an instance of a combinatorial optimization problem with $S^* \neq S$ and S finite. Furthermore, let N be a neighborhood function defined on S that induces a strongly connected neighborhood graph.*

If the generation matrix satisfies $G_{ij} = G_{ji}$ for all $i, j \in S$, then the finite homogeneous Markov chain associated with a run of simulated annealing at a fixed value c of the control parameter is strongly ergodic and the components of the unique stationary distribution $q(c)$ to which its probability distribution converges are given by

$$q_i(c) = \frac{1}{N_0(c)} \exp\left(-\frac{f(i)}{c}\right) \tag{8.7}$$

for all $i \in S$ with

$$N_0(c) = \sum_{j \in S} \exp\left(-\frac{f(j)}{c}\right).$$

Proof. By Theorem 8.4, it suffices to show that the stochastic vector $q(c)$ defined by (8.7) is a stationary distribution, i.e., a stochastic left eigenvector of $P(c)$ with eigenvalue 1. To this end, we only need to check whether for any $i, j \in S$ the 'detailed balance equation'

$$q_i(c)P_{ij}(c) = q_j(c)P_{ji}(c) \tag{8.8}$$

holds. This means that at equilibrium the probability that the Markov chain performs a transition from state i to state j is equal to the probability that it performs a transition from state j to state i. This can be seen as follows. If (8.8) holds, then

$$\sum_{j \in S} q_i(c)P_{ij}(c) = \sum_{j \in S} q_j(c)P_{ji}(c)$$

also holds. As $P(c)$ is stochastic, we have $\sum_{j \in S} P_{ij}(c) = 1$. This implies that

$q_i(c) = \sum_{j \in S} q_j(c) P_{ji}(c)$ and thus $q(c) = q(c) P(c)$, which means that q is a stationary distribution. Hence, it indeed suffices to prove the detailed balance equation (8.8) to conclude that $q(c)$ is a stationary distribution, which is to be proved.

The detailed balance equation trivially holds for $i = j$. Next, consider the case $i \neq j$. Since we have $G_{ij} = G_{ji}$, the detailed balance equation reduces then to $q_i(c) A_{ij}(c) = q_j(c) A_{ji}(c)$. Using the definition of $q(c)$ and $A(c)$, we derive

$$
\begin{aligned}
q_i(c) A_{ij}(c) &= \frac{1}{N_0(c)} \exp\left(-\frac{f(i)}{c}\right) \exp\left(-\frac{\max(f(j)-f(i),0)}{c}\right) \\
&= \frac{1}{N_0(c)} \exp\left(-\frac{f(j)}{c}\right) \exp\left(-\frac{f(i)-f(j)+\max(f(j)-f(i),0)}{c}\right) \\
&= \frac{1}{N_0(c)} \exp\left(-\frac{f(j)}{c}\right) \exp\left(-\frac{\max(f(i)-f(j),0)}{c}\right) \\
&= q_j(c) A_{ji}(c).
\end{aligned}
$$

\square

Example 8.3. Suppose that we apply simulated annealing to the neighborhood graph of Figure 8.2(a). As discussed in Section 7.1, simulated annealing can be viewed as a randomized version of threshold accepting, where the control parameter gives the expected value of the threshold. Theorem 8.5 yields that if the control parameter is 3, then the simulated annealing algorithm converges to the stationary distribution $q(3) \approx (0.38, 0.28, 0.20, 0.14)$. Hence, although the expected value of the threshold is as large as the maximum cost difference between any two solutions, the probability of ending up in the optimal solution is already 0.38. Furthermore, we get $q(1) \approx (0.64, 0.24, 0.09, 0.03)$ and $q(0.1) \approx (1, 5 \cdot 10^{-5}, 2 \cdot 10^{-9}, 9 \cdot 10^{-14})$. Hence, for $c = 0.1$, the probability of asymptotically ending up in a suboptimal solution is only $5 \cdot 10^{-5}$. \square

In the example we see that the smaller the fixed value of the control parameter, the smaller the probability that simulated annealing asymptotically ends up in a suboptimal solution. The following lemma states that this claim is generally true.

Lemma 8.2. *Let (S, f) be an instance of a combinatorial optimization problem. Furthermore, let $q(c)$ be given by (8.7), and let δ be the constant defined by $\min\{f(j) \mid j \in S \setminus S^*\} - f^*$, where f^* denotes the cost of an optimal solution. Then we have for any non-optimal solution $i \in S$*

$$
q_i(c) \leq \exp\left(-\frac{\delta}{c}\right). \tag{8.9}
$$

Proof. By (8.7), we know that for any solution i, either optimal or not, $q_i(c)$ is given by

$$
q_i(c) = \frac{\exp\left(-\frac{f(i)}{c}\right)}{\sum_{j \in S} \exp\left(-\frac{f(j)}{c}\right)} = \frac{\exp\left(\frac{f^*-f(i)}{c}\right)}{\sum_{j \in S} \exp\left(\frac{f^*-f(j)}{c}\right)}.
$$

As S contains at least one optimal solution, the sum in the denominator of the right-most expression is at least one. Hence, we have $q_i(c) \leq \exp\left(\frac{f^*-f(i)}{c}\right)$, from which the lemma follows. □

From Lemma 8.2 it follows that simulated annealing converges to the set of optimal solutions in the homogeneous model if the stationary distribution of the homogeneous Markov chain associated with a run of the algorithm at a fixed value c of the control parameter is given by (8.7).

Corollary 8.1. *Let (S, f) be an instance of a combinatorial optimization problem with $S^* \neq S$ and S finite. Furthermore, let N be a neighborhood function defined on S that induces a strongly connected neighborhood graph.*

If the generation matrix satisfies $G_{ij} = G_{ji}$ for all $i, j \in S$, then the finite homogeneous Markov chain associated with a run of simulated annealing at a fixed value c of the control parameter is strongly ergodic and the components of the unique stationary distribution $q(c)$ to which its probability distribution converges are given by (8.7). Furthermore, $q(c)$ satisfies

$$\lim_{c \downarrow 0} q_i(c) = 0 \ \text{for all} \ i \in S \setminus S^*.$$

□

From Lemma 8.2 one may suspect that an effective implementation of simulated annealing is obtained by assuming a very small, fixed value for the control parameter. This is, however, not the case. The reason is that (8.9) holds after an infinite number of iterations, while in practice the permissible number of iterations is relatively small. To substantiate this, observe that if the number of iterations exceeds the size of the solution space, then simulated annealing is less effective than total enumeration. Besides, if the control parameter has a low value and if we have only limited time, then the drawback that a large number of iterations is required to escape from a local optimum becomes eminent. To illustrate this, consider Example 8.2 and suppose that simulated annealing arrives in solution 2 in the neighborhood graph given in Figure 8.2(a). The random variable denoting the number of iterations required to leave this local optimum has the geometric distribution with parameter $P_{23}(c) + P_{24}(c)$, where a geometric random variable Y with parameter p has a probability distribution defined by $\mathbb{P}\{Y = k\} = p(1 - p)^{k-1}$ and an expected value of $\frac{1}{p}$. Because $P_{23}(c) = \frac{1}{2} \exp\left(-\frac{1}{c}\right)$ and $P_{24}(c) = \frac{1}{2} \exp\left(-\frac{2}{c}\right)$, this means that for $c = 0.1$ the expected number of iterations required to escape from solution 2 is as large as $50,000$. For $c = 3$ and $c = 1$, this expected number is only 1.6 and 4.0, respectively. This supports our claim that in an effective cooling schedule we want to start with a relatively large value for the control parameter.

In the remainder of this section we generalize the result of Corollary 8.1. Specifically, we weaken the condition that the generation matrix has to satisfy $G_{ij} = G_{ji}$ for all $i, j \in S$ to the condition that the neighborhood graph has to be symmetric.

Consider the stationary distribution $q(c)$ given by (8.7). Simple calculation shows that for any two solutions $i, j \in S$, the relation between the probabilities $q_i(c)$

and $q_j(c)$ is given by

$$q_i(c) = \exp\left(\frac{f(j) - f(i)}{c}\right) q_j(c).$$

Below, we prove for strongly connected, symmetric neighborhood graphs that a constant K exists, such that if we have $f(i) > f(j)$, then the more general relation

$$q_i(c) \leq K \exp\left(\frac{f(j) - f(i)}{c}\right) q_j(c)$$

holds. Based on this relation, we are next able to prove the claimed generalization of Corollary 8.1.

Lemma 8.3. *Let (S, f) be an instance of a combinatorial optimization problem with $S^* \neq S$ and S finite. Furthermore, let N be a neighborhood function defined on S that induces a strongly connected, symmetric neighborhood graph.*

The finite homogeneous Markov chain associated with a run of simulated annealing at a fixed value c of the control parameter is strongly ergodic. Furthermore, a constant K exists, such that for each control parameter value c and for each pair of solutions $i, j \in S$ with $f(i) > f(j)$ we have

$$q_i(c) \leq K \exp\left(\frac{f(j) - f(i)}{c}\right) q_j(c), \tag{8.10}$$

where $q(c)$ is the unique stationary distribution to which the probability distribution of the Markov chain converges.

Proof. By Definition 8.5 the generation matrix is independent of the value c of the control parameter. To prove the theorem, we prove the slightly more general result that it holds for generation matrices that may have some limited dependency on c. More precisely, we prove that the theorem holds if simulated annealing is based on a family $\{G(c) \mid c \in \mathbb{R}^+\}$ of generation matrices with the following two properties.

- Each generation matrix $G(c)$ satisfies Definition 8.5, which means that for each c and $i, j \in S$ we have $G_{ij}(c) > 0$ if and only if $j \in N(i)$.

- The 'ε-condition' is satisfied. This condition states that an $\varepsilon > 0$ exists such that for each c and $i, j \in S$ we have either $G_{ij}(c) = 0$ or $G_{ij}(c) > \varepsilon$.

We prove the generalized theorem by induction on the size of the solution space plus the number of distinct cost function values, i.e., on $m(S, f) = |S| + |\{f(s) \mid s \in S\}|$. For any $m(S, f)$, the existence of $q(c)$ follows from Theorem 8.4. Hence, what remains to be shown is that a constant K exists, such that for any c and $i, j \in S$ with $f(i) > f(j)$, the stationary distribution $q(c)$ satisfies (8.10).

Basis case. For $m(S, f) = 2$, we only have one single solution, which implies that the theorem trivially holds.

Induction step. Next, suppose that $m(S, f) \geq 3$. To prove that a constant K exists such that $q(c)$ satisfies (8.10) for each value $c > 0$, we distinguish the case that S contains two neighboring optimal solutions $s_1, s_2 \in S^*$ from the case that S does not contain such a pair of solutions.

Case 1: $s_1, s_2 \in S^$ with $s_2 \in N(s_1)$ exist.* We replace s_1 and s_2 by one single solution s'. In terms of the neighborhood graph of S, f, and N, the replacement of s_1 and s_2 corresponds to contracting the edge $\{s_1, s_2\}$. The cost of the new node is given by $f(s_1) = f(s_2)$. The obtained solution space, cost function, and neighborhood function are denoted by S', f', and N', respectively.

We now derive a family $\{G'(c) \mid c \in \mathbb{R}^+\}$ of generation matrices for S', f', N' from the assumed family $\{G(c) \mid c \in \mathbb{R}^+\}$ of generation matrices for S, f, N with the property that the unique stationary distributions $q'(c)$ and $q(c)$ of the corresponding Markov chains are related by

$$q'_i(c) = \begin{cases} q_i(c) & \text{if } i \neq s', \\ q_{s_1}(c) + q_{s_2}(c) & \text{if } i = s'. \end{cases} \tag{8.11}$$

Hence, asymptotically simulated annealing behaves the same on S, f, N as on S', f', N' except that in the latter case no distinction is made between nodes s_1 and s_2. It can be verified that $q'(c)$ satisfies (8.11) if we define $G'(c)$ as follows. If simulated annealing is at a solution $i \neq s'$, we generate a neighboring solution $j \in N'(i)$ with probability $G_{ij}(c)$ if $j \neq s'$ and with probability $G_{is_1}(c) + G_{is_2}(c)$ otherwise. If, on the other hand, we are at solution s', we consider this solution as solution s_1 with probability $q_{s_1}(c)/(q_{s_1}(c) + q_{s_2}(c))$ and as solution s_2 with probability $q_{s_2}(c)/(q_{s_1}(c) + q_{s_2}(c))$. Note that these probabilities are defined, i.e., $q_{s_1}(c) + q_{s_2}(c) > 0$ holds, because the strong connectivity of the neighborhood graph of S, f, N implies that $q_i > 0$ for all $i \in S$. We then generate neighboring solutions by using $G(c)$ accordingly. Formally, we define $G'(c)$ such that for any $i, j \in S'$

$$G'_{ij}(c) = \begin{cases} G_{ij}(c) & \text{if } i, j \neq s', \\ G_{is_1}(c) + G_{is_2}(c) & \text{if } i \neq s' \wedge j = s', \\ \frac{q_{s_1}(c)}{q_{s_1}(c) + q_{s_2}(c)} G_{s_1 j}(c) + \frac{q_{s_2}(c)}{q_{s_1}(c) + q_{s_2}(c)} G_{s_2 j}(c) & \text{if } i = s' \wedge j \neq s', \\ \frac{q_{s_1}(c)}{q_{s_1}(c) + q_{s_2}(c)} \left(G_{s_1 s_1}(c) + G_{s_1 s_2}(c) \right) & \\ \quad + \frac{q_{s_2}(c)}{q_{s_1}(c) + q_{s_2}(c)} \left(G_{s_2 s_2}(c) + G_{s_2 s_1}(c) \right) & \text{if } i = j = s'. \end{cases}$$

Obviously, the modified neighborhood graph corresponding to S', f', N' is still strongly connected and symmetric. Furthermore, we have $m(S', f') = m(S, f) - 1$: the size of the solution space has been reduced by one, and the number of distinct cost function values has not been changed. To apply the induction hypothesis, i.e., to assume that the generalized theorem holds for S', f', N', we still have to show for the family $\{G'(c) \mid c \in \mathbb{R}^+\}$ of generation matrices defined above that each generation matrix $G'(c)$ complies with Definition 8.5 and that the ε-condition is satisfied. Both requirements are by assumption satisfied by the family $\{G(c) \mid c \in \mathbb{R}^+\}$. Using this, it easily follows that the former requirement is also true for $\{G'(c) \mid c \in \mathbb{R}^+\}$. Furthermore, it implies that for proving the latter requirement it suffices to derive a positive lower bound on both $q_{s_1}(c)/(q_{s_1}(c) + q_{s_2}(c))$ and $q_{s_2}(c)/(q_{s_1}(c) + q_{s_2}(c))$ that does not depend on c. We now concentrate on this task.

As $f(s_1) = f(s_2)$ holds, which implies $A_{s_2 s_1}(c) = 1$, we have

$$q_{s_1}(c) = \sum_{i \in S} q_i(c) P_{i s_1}(c) \geq q_{s_2}(c) P_{s_2 s_1}(c) = q_{s_2}(c) G_{s_2 s_1}(c).$$

Since the ε-condition holds for matrix $G(c)$ and since $s_2 \in N(s_1)$, this yields that $q_{s_1}(c) \geq q_{s_2}(c)\varepsilon$. Hence, we obtain

$$\frac{q_{s_1}(c)}{q_{s_1}(c) + q_{s_2}(c)} \geq \frac{q_{s_1}(c)}{q_{s_1}(c) + \frac{1}{\varepsilon} q_{s_1}(c)} = \frac{\varepsilon}{1 + \varepsilon}.$$

Equivalently, it can be shown that $q_{s_2}(c)/(q_{s_1}(c) + q_{s_2}(c))$ is also bounded from below by $\frac{\varepsilon}{1+\varepsilon}$. This proves that the family $\{G'(c) \mid c \in \mathbb{R}^+\}$ satisfies the ε-condition. As a result, we can now apply the induction hypothesis to S', f', and N', i.e., we may assume that a K' exists, such that for each c and $i', j' \in S'$ with $f'(i') > f'(j')$ we have

$$q'_{i'}(c) \leq K' \exp\left(\frac{f'(j') - f'(i')}{c}\right) q'_{j'}(c), \tag{8.12}$$

where $q'(c)$ is the unique stationary distribution given by (8.11). We note that if $G(c)$ is independent of c, then this does not imply that $G'(c)$ is also independent of c. This is the reason that we prove a generalized version of the theorem.

To complete the proof of the induction hypothesis for the case that S contains the two neighboring optimal solutions s_1, s_2, we now show that (8.10) holds for all c and $i, j \in S$ with $f(i) > f(j)$ if K is defined as $\frac{1+\varepsilon}{\varepsilon} K'$.

First of all, consider the case that $j \notin \{s_1, s_2\}$. Note that $i \notin \{s_1, s_2\}$ holds because s_1 and s_2 are both optimal. Equation (8.10) follows when the definition of $q'(c)$ given by (8.11) is substituted in the equation obtained from (8.12) by taking $j' = j$ and $i' = i$.

Next, suppose that $j = s_1$. By taking $j' = s'$ and $i' = i$ in (8.12) and by applying the definition of $q'(c)$, we obtain

$$q_i(c) \leq K' \exp\left(\frac{f(s_1) - f(i)}{c}\right) (q_{s_1}(c) + q_{s_2}(c)).$$

Using the derived lower bound $\frac{\varepsilon}{1+\varepsilon}$ on $q_{s_1}(c)/(q_{s_1}(c) + q_{s_2}(c))$, we now get

$$q_i(c) \leq K' \exp\left(\frac{f(s_1) - f(i)}{c}\right) \frac{1+\varepsilon}{\varepsilon} q_{s_1}(c).$$

Hence, (8.10) holds for our choice of K. Similarly, this result can be derived for the case that $j = s_2$.

Case 2: $s_2 \notin N(s_1)$ for any $s_1, s_2 \in S^$.* To enable us to apply the induction hypothesis, we reduced the solution space S in Case 1 by merging two neighboring optimal solutions. In this case we reduce the other component in the definition of $m(S, f)$, namely the number of distinct cost function values. More specifically, we define a new cost function f' in which the cost of all optimal solutions is increased to the second-best cost. The increase is denoted by δ, which we have already defined in

Lemma 8.2. Hence, we have

$$\delta = \min\{f(i) \mid i \in S \setminus S^*\} - f^*$$

and $f'(i) = f(i) + \delta$ if $i \in S^*$ and $f'(i) = f(i)$ otherwise, where f^* denotes the cost of an optimal solution. We can now apply the induction hypothesis to S, f', N, and the assumed family $\{G(c) \mid c \in \mathbb{R}^+\}$ of generation matrices. This means that we may assume that a constant K' exists, such that for all c and $i, j \in S$ with $f'(i) > f'(j)$ the stationary distribution $q'(c)$ of the associated finite homogeneous Markov chain with transition matrix $P'(c)$ satisfies

$$q_i'(c) \le K' \exp\left(\frac{f'(j) - f'(i)}{c}\right) q_j'(c). \tag{8.13}$$

To use this equation to prove that the induction hypothesis also holds for S, f, and N, we express $q'(c)$ in terms of the stationary distribution $q(c)$ resulting from S, f, and N. To do this, we show that the vector $\hat{q}(c)$ with

$$\hat{q}_i(c) = \begin{cases} q_i(c) & \text{if } i \notin S^*, \\ \exp\left(-\frac{\delta}{c}\right) q_i(c) & \text{if } i \in S^* \end{cases}$$

is a left eigenvector of $P'(c)$ with eigenvalue 1. As vector $\hat{q}(c)$ is not stochastic, it does not define the stationary distribution $q'(c)$. However, $q'(c)$ is obtained from $\hat{q}(c)$ by normalizing it, i.e.,

$$q'(c) = \frac{\hat{q}(c)}{\sum_{i \in S} \hat{q}_i(c)}. \tag{8.14}$$

To prove that $\hat{q}(c)$ is a left eigenvector of $P(c)$, we derive two auxiliary results. Let $i, j \in S$ be arbitrary. If we do not have $i = j \in S^*$, then

$$P_{ij}'(c)\hat{q}_i(c) = P_{ij}(c)q_i(c) \tag{8.15}$$

holds. This is obviously true if $i \notin S^*$ as in that case we have $\hat{q}_i(c) = q_i(c)$ and $P_{ij}'(c) = P_{ij}(c)$. Next, suppose that $i \in S^*$. If in addition $j \in S^*$, then (8.15) follows from $P_{ij}'(c) = P_{ij}(c) = 0$, which is implied by the assumption that for any two different optimal solutions s_1 and s_2 we have $s_2 \notin N(s_1)$. If, on the other hand, $j \notin S^*$ holds, then we have $\hat{q}_i(c) = \exp\left(-\frac{\delta}{c}\right) q_i(c)$ and, as $A_{ij}'(c) = \exp\left(\frac{\delta}{c}\right) A_{ij}(c)$, also $P_{ij}'(c) = \exp\left(\frac{\delta}{c}\right) P_{ij}(c)$. This again yields (8.15).

Next, suppose that $i = j \in S^*$. By definition, we then have

$$P_{jj}'(c)\hat{q}_j(c) = \left(1 - \sum_{l \in S, l \ne j} P_{jl}'(c)\right) \hat{q}_j(c).$$

By (8.15), the right-hand side can be rewritten to

$$\hat{q}_j(c) - \sum_{l \in S, l \ne j} P_{jl}(c)q_j(c)$$

and, as $\sum_{l \in S} P_{jl} = 1$, this gives

$$
\begin{aligned}
P'_{jj}(c)\hat{q}_j(c) &= \hat{q}_j(c) - (1 - P_{jj}(c))q_j(c) \\
&= P_{jj}(c)q_j(c) - q_j(c) + \hat{q}_j(c). \qquad (8.16)
\end{aligned}
$$

Using (8.15), (8.16), and the observation that $q(c)$ is a left eigenvector of $P(c)$, we can now derive

$$
\sum_{i \in S} P'_{ij}(c)\hat{q}_i(c) = \sum_{i \in S} P_{ij}(c)q_i(c) - q_j(c) + \hat{q}_j(c) = \hat{q}_j(c).
$$

Hence, $\hat{q}(c)$ is indeed a left eigenvector of $P'(c)$, which implies that $q'(c)$ given by (8.14) is a stationary distribution of the Markov chain associated with $P'(c)$.

Let $z(c) = \sum_{i \in S} \hat{q}_i(c)$ be the normalization factor in (8.14). Then, if we substitute the definitions of $q'(c)$ and f' in (8.13), we obtain that for any c and $i, j \in S$ with $f(i) > f(j)$ we have

$$
\frac{q_i(c)}{z(c)} \leq K' \exp\left(\frac{f(j) - f(i)}{c}\right) \frac{q_j(c)}{z(c)}
$$

if $j \notin S^*$ and

$$
\frac{q_i(c)}{z(c)} \leq K' \exp\left(\frac{f(j) + \delta - f(i)}{c}\right) \frac{\exp\left(-\frac{\delta}{c}\right) q_j(c)}{z(c)}
$$

if $j \in S^*$. This implies that (8.10) holds for $K = K'$ and for all c and $i, j \in S$ with $f(i) > f(j)$, which completes the proof of the induction hypothesis. □

Using Lemma 8.3 we are able to generalize Lemma 8.2 and, consequently, Corollary 8.1.

Lemma 8.4. *Let (S, f) be an instance of a combinatorial optimization problem. Furthermore, let stochastic vector $q(c)$ satisfy (8.10) for a constant K and each pair of solutions $i, j \in S$ with $f(i) > f(j)$, and let δ be the constant defined by $\min\{f(i) \mid i \in S \setminus S^*\} - f^*$, where f^* denotes the cost of an optimal solution. Then we have for any non-optimal solution $i \in S$*

$$
q_i(c) \leq K \exp\left(-\frac{\delta}{c}\right).
$$

Proof. Because $q_j(c) \leq 1$ for all $j \in S$, (8.10) implies that $q_i \leq K \exp\left(\frac{f(j) - f(i)}{c}\right)$ for any j with $f(i) > f(j)$. By letting j be an optimal solution, the lemma follows. □

Corollary 8.2. *Let (S, f) be an instance of a combinatorial optimization problem with $S^* \neq S$ and S finite. Furthermore, let N be a neighborhood function defined on S that induces a strongly connected, symmetric neighborhood graph.*

Then the finite homogeneous Markov chain associated with a run of simulated annealing at a fixed value c of the control parameter is strongly ergodic and the unique stationary distribution $q(c)$ to which its probability distribution converges

satisfies

$$\lim_{c \downarrow 0} q_i(c) = 0$$

for any non-optimal solution $i \in S$. □

8.4 Equilibrium Statistics

We now derive some statistical results for simulated annealing in the model discussed in the previous section, i.e., in the case that we lower the control parameter only if equilibrium is attained. We will assume that all conditions of Theorem 8.5 are satisfied, where we draw special attention to the condition that the generation matrix has to satisfy $G_{ij} = G_{ji}$ for all $i, j \in S$.

By Theorem 8.5, we know that, at any value c of the control parameter, the probability of being in solution $i \in S$ at equilibrium is given by (8.7), i.e., by $q_i(c) = \frac{1}{N_0(c)} \exp\left(-\frac{f(i)}{c}\right)$, where $N_0(c) = \sum_{j \in S} \exp\left(-\frac{f(j)}{c}\right)$. Hence, at equilibrium the expected cost $\langle f \rangle_c$ and the variance σ_c^2 of the cost are given by

$$\langle f \rangle_c = \sum_{i \in S} f(i) q_i(c) \tag{8.17}$$

and

$$\sigma_c^2 = \sum_{i \in S} (f(i) - \langle f \rangle_c)^2 q_i(c), \tag{8.18}$$

respectively. Note that because $S \neq S^*$ by one of the conditions of Theorem 8.5 and because $q_i(c) > 0$ for all $c > 0$, we have $\sigma_c^2 > 0$. Using this observation and the following theorem yields that if the control parameter decreases, then the expected cost at equilibrium also decreases.

Theorem 8.6. *Let* $\langle f \rangle_c$ *and* σ_c^2 *be defined by (8.17) and (8.18), respectively. Then, we have*

$$\frac{\partial}{\partial c} \langle f \rangle_c = \frac{\sigma_c^2}{c^2}.$$

Proof. Using the definition of $\langle f \rangle_c$ gives

$$\frac{\partial}{\partial c} \langle f \rangle_c = \sum_{i \in S} f(i) \frac{\partial}{\partial c} q_i(c). \tag{8.19}$$

Furthermore, simple calculus yields

$$\frac{\partial}{\partial c} N_0(c) = \sum_{j \in S} \frac{f(j)}{c^2} \exp\left(\frac{-f(j)}{c}\right)$$

and thus, by using $\frac{\partial}{\partial c}(g(c) \cdot h(c)) = g(c)\frac{\partial}{\partial c}h(c) + h(c)\frac{\partial}{\partial c}g(c)$ with $g(c) = 1/N_0(c)$ and $h(c) = \exp\left(\frac{-f(i)}{c}\right)$, we obtain

$$
\begin{aligned}
\frac{\partial}{\partial c} q_i(c) &= \frac{\partial}{\partial c} \frac{\exp\left(\frac{-f(i)}{c}\right)}{N_0(c)} \\
&= \frac{f(i)}{c^2} \frac{\exp\left(\frac{-f(i)}{c}\right)}{N_0(c)} - \frac{\exp\left(\frac{-f(i)}{c}\right)}{N_0^2(c)} \frac{\partial}{\partial c} N_0(c) \\
&= \frac{q_i(c)}{c^2} f(i) - \frac{q_i(c)}{c^2} \frac{\sum_{j \in S} f(j) \exp\left(\frac{-f(j)}{c}\right)}{N_0(c)} \\
&= \frac{q_i(c)}{c^2} (f(i) - \langle f \rangle_c).
\end{aligned}
$$

Substituting this for $\frac{\partial}{\partial c} q_i(c)$ in (8.19) now gives

$$
\begin{aligned}
\frac{\partial}{\partial c} \langle f \rangle_c &= \frac{1}{c^2} \sum_{i \in S} q_i(c) \left(f^2(i) - f(i) \langle f \rangle_c \right) \\
&= \frac{1}{c^2} \left(-\langle f \rangle_c^2 + \sum_{i \in S} f^2(i) q_i(c) \right).
\end{aligned}
$$

By definition, we have $\langle f \rangle_c = \sum_{i \in S} f(i) q_i(c)$. Furthermore, because $\sum_{i \in S} q_i(c) = 1$ we also have $\langle f \rangle_c = \sum_{i \in S} \langle f \rangle_c q_i(c)$. Using these observations we obtain

$$
\begin{aligned}
\frac{\partial}{\partial c} \langle f \rangle_c &= \frac{1}{c^2} \left(-2\langle f \rangle_c^2 + \langle f \rangle_c^2 + \sum_{i \in S} f^2(i) q_i(c) \right) \\
&= \frac{1}{c^2} \sum_{i \in S} (f^2(i) - 2f(i)\langle f \rangle_c + \langle f \rangle_c^2) q_i(c) \\
&= \frac{1}{c^2} \sum_{i \in S} (f(i) - \langle f \rangle_c)^2 q_i(c) \\
&= \frac{\sigma_c^2}{c^2}.
\end{aligned}
$$

\square

By using the following theorem, we can strengthen the observation that the expected cost at equilibrium decreases if the control parameter decreases. It implies that the expected cost at equilibrium decreases from a value that is at most the average cost to the optimal cost.

Theorem 8.7. *Let $\langle f \rangle_c$ and σ_c^2 be defined by (8.17) and (8.18), respectively. Furthermore, let $\langle f \rangle_\infty = \frac{1}{|S|} \sum_{i \in S} f(i)$ denote the average cost and let f^* denote the*

optimal cost. Then, we have

$$\lim_{c \to \infty} \langle f \rangle_c = \langle f \rangle_\infty,$$

$$\lim_{c \downarrow 0} \langle f \rangle_c = f^*,$$

$$\lim_{c \to \infty} \sigma_c^2 = \frac{1}{|S|} \sum_{i \in S} (f(i) - \langle f \rangle_\infty)^2, \text{ and}$$

$$\lim_{c \downarrow 0} \sigma_c^2 = 0.$$

Proof. The theorem can be proved by using the definitions of $\langle f \rangle_c$, σ_c^2, and $q(c)$. \square

In the above two theorems we studied the solution space as a whole. For an individual solution $i \in S$, we can derive the following result. For decreasing c, the probability of being in solution i at equilibrium increases if i is optimal and it decreases if i has a cost that is at least the average cost. Otherwise, a threshold c_i exists, such that the probability of being in solution i increases if $c > c_i$ and decreases if $c < c_i$.

Theorem 8.8. *Let $\langle f \rangle_\infty = \frac{1}{|S|} \sum_{i \in S} f(i)$ denote the average cost, and let f^* denote the optimal cost. Furthermore, let the components of distribution $q(c)$ of S be given by (8.7). If $S \neq S^*$, then we have for any $i \in S$*

$$\frac{\partial}{\partial c} q_i(c) < 0$$

if $f(i) = f^$,*

$$\frac{\partial}{\partial c} q_i(c) > 0$$

if $f(i) \geq \langle f \rangle_\infty$, and

$$\frac{\partial}{\partial c} q_i(c) \begin{cases} < 0 & \text{if } c > c_i \\ = 0 & \text{if } c = c_i \\ > 0 & \text{if } c < c_i \end{cases}$$

for a $c_i > 0$ if $f^ < f(i) < \langle f \rangle_\infty$.*

Proof. In Theorem 8.6 we proved

$$\frac{\partial}{\partial c} q_i(c) = \frac{q_i(c)}{c^2} (f(i) - \langle f \rangle_c).$$

As $\frac{q_i(c)}{c^2} > 0$, this implies that the sign of $\frac{\partial}{\partial c} q_i(c)$ is determined by the sign of $f(i) - \langle f \rangle_c$. As we concluded from Theorems 8.6 and 8.7, $\langle f \rangle_c$ increases from f^* to $\langle f \rangle_\infty$ when c increases. The theorem now follows easily. \square

8.5 *Inhomogeneous Markov Chains

In Section 8.3 we proved that simulated annealing converges to the set of optimal solutions in the unrealistic setting that the control parameter is only changed after equilibrium has been attained. The main result of this section is that convergence in

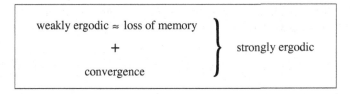

Figure 8.5. Definition of strong ergodicity.

this unrealistic homogeneous model implies convergence of a real run of simulated annealing if we apply a cooling schedule in which the control parameter is lowered slowly enough.

Before making the result more precise, we summarize the results we derived for the homogeneous model. We first showed in Section 8.2 that if the neighborhood graph is strongly connected, then the homogeneous Markov chain associated with a run of simulated annealing at some fixed value c of the control parameter is strongly ergodic, which means that the probability distribution of the Markov chain converges to a unique stationary distribution $q(c)$, regardless of how a starting solution is obtained. Next, we derived in Section 8.3 that, under the additional assumption that the neighborhood graph is symmetric, we have $\lim_{c \downarrow 0} q_i(c) = 0$ for any suboptimal solution $i \in S$, i.e., the probability of being in a suboptimal solution at equilibrium becomes zero as c approaches zero.

In this section we show that if the neighborhood graph is strongly connected, then the single inhomogeneous Markov chain associated with a real run of simulated annealing is also strongly ergodic and the stochastic vector to which its probability distribution converges is given by $q = \lim_{c \downarrow 0} q(c)$, where $q(c)$ is as defined above. This result confirms our claim that convergence in the homogeneous model implies convergence in a realistic setting. Note, however, that the result holds under the assumption that the neighborhood graph is strongly connected. Hence, suppose that we can prove for the homogeneous model that simulated annealing still converges to the set of optimal solutions after we have replaced the condition on the neighborhood graph that stipulates that it must be symmetric with a different condition. In that case the result implies that a real run of simulated annealing also converges to the set of optimal solutions. If, on the other hand, we are able to prove for the homogeneous model that simulated annealing converges to the set of optimal solutions for a type of neighborhood graph that is not strongly connected (but, among other properties, is weakly optimally connected), then the result does not imply that a real run of simulated annealing also converges to the set of optimal solutions.

To prove our result, we need some general Markov chain theory on how to prove that an inhomogeneous Markov chain is strongly ergodic. For a Markov chain the characteristic of being strongly ergodic implies two things: loss of memory and convergence, where loss of memory means that the effect of the initial probability distribution of its states is lost; see Figure 8.5. If a Markov chain shows a loss of memory, but does not necessarily converge, then it is said to be weakly ergodic.

Definition 8.13. A Markov chain is called *weakly ergodic* if for all $t \geq 0$ and each triple i, j, l of states we have

$$\lim_{t' \to \infty} \left(\mathbb{P}\{X(t') = i \mid X(t) = j\} - \mathbb{P}\{X(t') = i \mid X(t) = l\} \right) = 0. \qquad (8.20)$$

\square

The definition of weak ergodicity is slightly stronger than required to enforce loss of memory. Loss of memory is already guaranteed if (8.20) holds for $t = 0$. However, we do not want the loss of memory to be realized only by the first few transitions of the Markov chain, which would imply that the outcomes of these early transitions have an essential effect on the behavior of the Markov chain in the long run. Instead, we prefer that for any $t \geq 0$ the effect of the outcomes of the first t transitions vanishes as time proceeds, i.e., we want the memory to be lost independently of when the process started. This clarifies our definition of weak ergodicity. For similar reasons, we defined strong ergodicity in Definition 8.10, such that (8.5) has to hold for any $t \geq 0$ instead of for only $t = 0$.

For finite homogeneous Markov chains there is no distinction between weak and strong ergodicity, as is stated by the following theorem.

Theorem 8.9. *If a finite homogeneous Markov chain is weakly ergodic, then it is also strongly ergodic.*

Proof. Consider an arbitrary finite homogeneous Markov chain, and assume that it is weakly ergodic. The weak ergodicity implies that the probability distribution $a(t')$ of $X(t')$ for $t' \to \infty$ does not depend on the probability distribution $a(t)$ of any $X(t)$ with $t \geq 0$ fixed. Furthermore, the Markov chain has at least one stationary distribution q by Theorem 8.1, which means that if $X(t)$ has probability distribution q, then all random variables $X(t')$ with $t' \geq t$ have probability distribution q. By combining these two observations we obtain that $a(t')$ approaches q as $t' \to \infty$, regardless of $a(t)$. This proves the theorem. \square

For inhomogeneous Markov chains, weak ergodicity need not imply strong ergodicity, as follows from the following example.

Example 8.4. Consider the inhomogeneous Markov chain with state space $\{1,2\}$ and transition matrix

$$P(t) = \begin{pmatrix} 1 - \frac{1}{t} & \frac{1}{t} \\ \frac{1}{t} & 1 - \frac{1}{t} \end{pmatrix}$$

if t is odd and

$$P(t) = \begin{pmatrix} \frac{1}{t} & 1 - \frac{1}{t} \\ 1 - \frac{1}{t} & \frac{1}{t} \end{pmatrix}$$

if t is even. The corresponding transition graphs are depicted in Figure 8.6. If t is odd, then, regardless of the current state of the Markov chain, the probability of going to state 1 is given by $1 - \frac{1}{t}$ and the probability of going to state 2 by $\frac{1}{t}$. Similarly, if t is even, then the probability of going to state 1 is given by $\frac{1}{t}$ and the probability of going to state 2 by $1 - \frac{1}{t}$. Hence, we have $a(t) = (1 - \frac{1}{t}, \frac{1}{t})$ for

t odd t even

Figure 8.6. State transition graphs corresponding to the two transition matrices of the inhomogeneous Markov chain of Example 8.4.

t odd and $a(t) = (\frac{1}{t}, 1 - \frac{1}{t})$ for t even. As this result holds independently of $a(t')$ with $t' < t$, this shows that the Markov chain is weakly ergodic. However, it is not strongly ergodic as it does not converge: for $t \to \infty$ the probability distribution $a(t)$ alternates between $(1,0)$ and $(0,1)$. □

From the example above it follows that, although it is a necessary condition, weak ergodicity alone does not guarantee strong ergodicity for an inhomogeneous Markov chain. A possible condition that, in combination with weak ergodicity, guarantees strong ergodicity is given by the following theorem. The condition is expressed in terms of the left eigenvectors with eigenvalue 1 of the transition matrices of the Markov chain, i.e., in terms of the stationary distributions of the homogeneous Markov chains associated with the successive, possibly different, transition matrices of the inhomogeneous Markov chain. By Theorem 8.1 at least one eigenvector with eigenvalue 1 exists for each transition matrix $P(t)$. The additional condition implies that the stationary distributions of the homogeneous Markov chains associated with the successive transition matrices of the inhomogeneous Markov chain converge to a distribution, where an arbitrary stationary distribution of a homogeneous Markov chain may be selected if there are multiple alternatives. In the theorem we use the notation $||v||$ to denote the norm of vector v, i.e., $||v|| = \sum_i |v_i|$.

Theorem 8.10. *A finite inhomogeneous Markov chain is strongly ergodic under the following two conditions.*

- *The Markov chain is weakly ergodic and*
- *a sequence $\{q(t)\}_{t=1}^{\infty}$ of eigenvectors with eigenvalue 1 exists, one for each transition matrix $P(t)$, such that*

$$\sum_{t=1}^{\infty} ||q(t) - q(t+1)|| < \infty.$$

Moreover, the stochastic vector to which the probability distribution $a(t)$ converges for $t \to \infty$ is given by $q = \lim_{t \to \infty} q(t)$.

Proof. See the proof of Theorem V.4.3 in Isaacson & Madsen [1976]. □

Using this theorem, we will be able to prove our main result, i.e., that for a strongly connected neighborhood graph a cooling schedule exists such that simulated annealing implies a strongly ergodic inhomogeneous Markov chain for which the proba-

bility distribution converges to $q = \lim_{c \downarrow 0} q(c)$, where $q(c)$ is the unique stationary distribution of the inhomogeneous Markov chain associated with a run of simulated annealing at a fixed value c of the control parameter.

In order to apply the theorem we still have to find an approach to show that an inhomogeneous Markov chain is weakly ergodic, since it is generally difficult to use its definition directly, as we did in Example 8.4. We first focus on this problem.

By definition, a Markov chain is weakly ergodic if for any pair $a(t), a'(t)$ of probability distributions we have

$$\lim_{t' \to \infty} \left(a(t) \prod_{i=t}^{t'} P(i) - a'(t) \prod_{i=t}^{t'} P(i) \right) = 0.$$

In other words, $a(t) \prod_{i=t}^{t'} P(i)$ must become independent of $a(t)$ as $t' \to \infty$. It can be verified that $a(t) \prod_{i=t}^{t'} P(i)$ is independent of $a(t)$ for some fixed t' if and only if $\prod_{i=t}^{t'} P(i)$ is a constant matrix, where a matrix is said to be constant if all its rows are identical. This suggests that the weak ergodicity of an inhomogeneous Markov chain depends on whether $\prod_{i=t}^{t'} P(i)$ approaches the constant matrix for $t' \to \infty$. To formalize the idea of approaching a constant matrix, we use the ergodic coefficient.

Definition 8.14. Let P be a $n \times n$ stochastic matrix. Then the *ergodic coefficient* $\alpha(P)$ is defined as

$$\alpha(P) = \min_{1 \le i \le j \le n} \sum_{l=1}^{n} \min(P_{il}, P_{jl})$$

or, equivalently, as

$$\alpha(P) = 1 - \frac{1}{2} \max_{1 \le i \le j \le n} \sum_{l=1}^{n} |P_{il} - P_{jl}|.$$

\square

For a proof that the two definitions that are given for the ergodic coefficient $\alpha(P)$ of a stochastic matrix P are indeed equivalent, we refer to Exercise 3 at the end of this chapter. It can be verified that the ergodic coefficient $\alpha(P)$ is a value between zero and the maximum sum of the entries in a row, which is one. Furthermore, the coefficient equals one if and only if P is a constant matrix and it equals zero if and only if P contains two rows with the property that a positive entry in one row implies a zero entry in the other row. In terms of an inhomogeneous Markov chain, this means that if the ergodic coefficient of $\prod_{i=t}^{t'} P(i)$ equals one, then the probability of being in state l at time t' is independent of the state at time t. If, on the other hand, the ergodic coefficient equals zero, then two states i and j exist, such that the set of possible states at time t' when being in state i at time t is disjunct from the possible states at time t' when being in state j at time t.

We can now formalize the property of approaching a constant matrix as the property that the ergodic coefficient approaches one. Using this formalization, the assumption given above that an inhomogeneous Markov chain is weakly ergodic if

and only if $\prod_{i=t}^{t'} P(i)$ approaches the constant matrix for $t' \to \infty$ can be proved to be correct.

Theorem 8.11. *A finite inhomogeneous Markov chain is weakly ergodic if and only if for all $t \geq 1$ we have*

$$\lim_{t' \to \infty} \alpha \left(\prod_{i=t}^{t'} P(i) \right) = 1.$$

Proof. See Theorem V.3.1 in Isaacson & Madsen [1976]. □

Based on this result, we can prove an alternative characterization of weak ergodicity. This is the characterization that we will use.

Theorem 8.12. *A finite inhomogeneous Markov chain is weakly ergodic if and only if an increasing sequence $\{n_i\}_{i=0}^{\infty}$ with $n_0 = 0$ exists, such that*

$$\sum_{i=0}^{\infty} \alpha \left(\prod_{t=n_i+1}^{n_{i+1}} P(t) \right) = \infty.$$

Proof. See Theorem V.3.2 in Isaacson & Madsen [1976]. □

This concludes our discussion on general Markov chain theory. We now apply this theory to show that if the neighborhood graph is strongly connected and the control parameter is lowered slowly enough, then the inhomogeneous Markov chain associated with a run of simulated annealing is strongly ergodic and converges in distribution to $q = \lim_{c \downarrow 0} q(c)$. Following the sufficient conditions of Theorem 8.10, we first prove Lemma 8.7, which states that the inhomogeneous Markov chain is weakly ergodic. In this proof we write $\underline{G} = \min_{i \in S, j \in N(i)} G_{ij}$ to denote the smallest generation probability and $\underline{A}(c) = \min_{i \in S, j \in N(i)} A_{ij}(c)$ to denote the smallest probability of a neighboring solution being accepted. Note that for a strongly connected neighborhood graph with $S^* \neq S$ we have $0 < \underline{A}(c) < 1$.

Obviously, the probability $A_{ij}(c)$ of any cost-deteriorating neighbor being accepted converges to zero as $c \downarrow 0$. Lemma 8.6 states that the inhomogeneous Markov chain associated with simulated annealing is weakly ergodic if this convergence is not too fast. To prove Lemma 8.6, the following lemma is useful.

Lemma 8.5. *Consider the inhomogeneous Markov chain associated with a run of simulated annealing for a sequence $\{c_k\}_{k=1}^{\infty}$ of control parameter values. If a solution i has at least one neighbor with cost strictly larger than the cost of i and if $c_k \leq \ln^{-1} 2$, then we have $P_{ii}(c_k) \geq \underline{G} \cdot \underline{A}(c_k)$.*

Proof. Let j be a neighbor of i with strictly higher cost. By definition we have

$$P_{ii}(c_k) = 1 - \sum_{l \in S, l \neq i} G_{il} A_{il}(c_k) = 1 - G_{ij} A_{ij}(c_k) - \sum_{l \in S, l \neq i,j} G_{il} A_{il}(c_k).$$

Since $A_{il} \leq 1$ for all $l \in S$ and $\sum_{l \in S} G_{il} = 1$, we now get

$$P_{ii}(c_k) \geq 1 - G_{ij} A_{ij}(c_k) - (1 - G_{ij}) \geq (1 - A_{ij}(c_k)) G_{ij}.$$

Because $f(i) < f(j)$ holds, we have by definition $A_{ij}(c_k) = \exp\left(\frac{f(i)-f(j)}{c_k}\right)$. If we use this and $c_k \leq \ln^{-1} 2$, calculus yields $1 - A_{ij}(c_k) \geq A_{ij}(c_k)$. Hence,

$$P_{ii}(c_k) \geq G_{ij}A_{ij}(c_k) \geq \underline{G} \cdot \underline{A}(c_k)$$

holds, which proves the lemma. □

Lemma 8.6. *Let (S,f) be an instance of a combinatorial optimization problem with $S^* \neq S$ and S finite. Furthermore, let N be a neighborhood function defined on S that induces a strongly connected neighborhood graph with diameter d.*

If we assume a cooling schedule in which the sequence $\{c_k\}_{k=1}^{\infty}$ of control parameter values is non-increasing, starts with value $c_1 \leq \ln^{-1} 2$, and satisfies $\sum_{i=1}^{\infty} \underline{A}^d(c_{i \cdot d}) = \infty$, then the inhomogeneous Markov chain associated with a run of simulated annealing is weakly ergodic.

Proof. To prove the lemma, we show that the ergodic coefficient of $\prod_{k=(i-1)d+1}^{i \cdot d} P(c_k)$ is at least $\underline{G}^d \underline{A}^d(c_{i \cdot d})$ for any $i \geq 1$. In this case the assumption $\sum_{i=1}^{\infty} \underline{A}^d(c_{i \cdot d}) = \infty$ yields

$$\sum_{i=1}^{\infty} \alpha \left(\prod_{k=(i-1)d+1}^{i \cdot d} P(c_k) \right) \geq \underline{G}^d \sum_{i=1}^{\infty} \underline{A}^d(c_{i \cdot d}) = \infty,$$

which by Theorem 8.12 implies that an inhomogeneous Markov chain associated with simulated annealing is indeed weakly ergodic.

Let $P^{(i)}$ be a shorthand notation for the transition matrix $\prod_{k=(i-1)d+1}^{i \cdot d} P(c_k)$. Our proof obligation can now be formalized as $\alpha(P^{(i)}) \geq \underline{G}^d \underline{A}^d(c_{i \cdot d})$ for all $i \geq 1$. By definition, we have

$$\alpha \left(P^{(i)} \right) = \min_{j,j' \in S} \sum_{l \in S} \min(P_{jl}^{(i)}, P_{j'l}^{(i)}) \geq \min_{j \in S} P_{js^*}^{(i)},$$

where s^* is an optimal solution. Hence, it now suffices to derive $P_{js^*}^{(i)} \geq \underline{G}^d \underline{A}^d(c_{i \cdot d})$ for any $i \geq 1$ and $j \in S$.

As the neighborhood graph has diameter $d < \infty$, it contains a path p that successively visits the solutions s_1, s_2, \ldots, s_n with $s_1 = j$, $s_n = s^*$, and $n \leq d$. By the definition of $P^{(i)}$, we have

$$P_{js^*}^{(i)} \geq \left(\prod_{k=(i-1)d+1}^{(i-1)d+n-1} P(c_k) \right)_{j,s^*} + \left(\prod_{k=(i-1)d+n}^{i \cdot d} P(c_k) \right)_{s^*,s^*}. \quad (8.21)$$

The first product is at least $\prod_{k=(i-1)d+1}^{(i-1)d+n-1} \underline{G} \cdot \underline{A}(c_k)$, as follows from the definitions of \underline{G} and \underline{A} and the existence of path p. Next, consider the second product. As $S^* \neq S$ and the neighborhood graph is strongly connected, solution s^* contains at least one neighbor with higher cost. Furthermore, $c_k \leq \ln^{-1} 2$ as, by assumption, $c_1 \leq \ln^{-1} 2$ and $\{c_k\}_{k=1}^{\infty}$ is non-increasing. By Lemma 8.5 this implies that $P_{s^*s^*}(c_k) \geq \underline{G} \cdot \underline{A}(c_k)$ for any k. Hence, the second product in (8.21) is at least $\prod_{k=(i-1)d+n}^{i \cdot d} \underline{G} \cdot \underline{A}(c_k)$.

Combining these results gives

$$P_{js^*}^{(i)} \geq \prod_{k=(i-1)d+1}^{i \cdot d} \underline{G} \cdot \underline{A}(c_k).$$

As c_k, and consequently $\underline{A}(c_k)$, is non-increasing in k, by assumption, the right-hand side expression is at least $\underline{G}^d \underline{A}^d(c_{i \cdot d})$, which was what we wanted to prove. □

Lemma 8.7. *Let (S, f) be an instance of a combinatorial optimization problem with $S^* \neq S$ and S finite. Furthermore, let N be a neighborhood function defined on S that induces a strongly connected neighborhood graph with diameter d.*

If we assume a cooling schedule for which a k_0 exists, such that the sequence $\{c_k\}_{k=1}^{\infty}$ of control parameter values is non-increasing and satisfies

$$\frac{d\Delta}{\log k} \leq c_k \leq \frac{1}{\ln 2} \tag{8.22}$$

for all $k \geq k_0$ with $\Delta = \max_{i \in S, j \in N(i)} (f(j) - f(i))$, then the inhomogeneous Markov chain associated with a run of simulated annealing is weakly ergodic.

Proof. If the inhomogeneous Markov chain is weakly ergodic in the case that we switch to a cooling schedule in which the sequence of control parameter values starts at k_0 instead of c_1, then it is also weakly ergodic for the original sequence $\{c_k\}_{k=1}^{\infty}$, as follows from the definition of weak ergodicity. Hence, by Lemma 8.6 it suffices to show that $\sum_{i=k_0}^{\infty} \underline{A}^d(c_{i \cdot d}) = \infty$ holds. By definition, we have $\underline{A}(c) = \exp\left(-\frac{\Delta}{c}\right)$. Hence, by using (8.22) we can derive

$$\sum_{i=k_0}^{\infty} \underline{A}^d(c_{i \cdot d}) \geq \sum_{i=k_0}^{\infty} \exp\left(-d\frac{\Delta}{d\Delta/\log(id)}\right)$$

$$= \sum_{i=k_0}^{\infty} \exp\left(-\log(id)\right)$$

$$= \frac{1}{d} \sum_{i=k_0}^{\infty} \frac{1}{i},$$

which is ∞. □

Theorem 8.13. *Let (S, f) be an instance of a combinatorial optimization problem with $S^* \neq S$ and S finite. Furthermore, let N be a neighborhood function defined on S that induces a strongly connected neighborhood graph with diameter d.*

If a cooling schedule is assumed for which a k_0 exists, such that the sequence $\{c_k\}_{k=1}^{\infty}$ of control parameter values is non-increasing and satisfies

$$\frac{d\Delta}{\log k} \leq c_k \leq \frac{1}{\ln 2}$$

for all $k \geq k_0$ with $\Delta = \max_{i \in S, j \in N(i)} (f(j) - f(i))$, then the inhomogeneous Markov chain associated with a run of simulated annealing is strongly ergodic and its probability distribution converges to $q = \lim_{k \to \infty} q(c_k)$, where $q(c_k)$ is the unique sta-

tionary distribution of the homogeneous Markov chain associated with a run of simulated annealing at a fixed value c_k of the control parameter.

Proof. By Theorem 8.4 we have that for any $c > 0$ the homogeneous Markov chain associated with simulated annealing has a unique stationary distribution $q(c)$ when run at a fixed value c of the control parameter. By definition, this means that for each k transition matrix $P(c_k)$ has $q(c_k)$ as its unique left eigenvector with eigenvalue 1. If we now invoke Theorem 8.10 and Lemma 8.7 this means that to prove the theorem it suffices to show that

$$\sum_{k=1}^{\infty} \|q(c_k) - q(c_{k+1})\| < \infty.$$

Because S is finite, this is true if for each $i \in S$ we have

$$\sum_{k=1}^{\infty} |q_i(c_k) - q_i(c_{k+1})| < \infty. \tag{8.23}$$

To prove this, we first show that $q_i(c)$ is a continuously differentiable function on the open interval $(0, \infty)$.

The stochastic vector $q(c)$ is the only feasible solution in the polytope defined by

$$q(c)(P(c) - I) = \overline{0} \tag{8.24}$$

$$\sum_{i \in S} q_i(c) = 1 \tag{8.25}$$

$$q(c) \geq \overline{0}, \tag{8.26}$$

where I denotes the $|S| \times |S|$ identity matrix. The linear constraints given by (8.24) can be rewritten to $q(c) = q(c)P(c)$ and thus ensure that $q(c)$ is a left eigenvector of $P(c)$ with eigenvalue 1. The constraints given by (8.25) and (8.26) ensure that $q(c)$ is stochastic. From the linear constraints, we can safely remove any of the $|S|$ constraints of (8.24), i.e., we can remove $\sum_{i \in S} q_i(c)(P(c) - I)_{ij} = 0$ for any single $j \in S$. This can be seen as follows. Suppose that for all $l \in S$ with $l \neq j$, we have $\sum_{i \in S} q_i(c)(P(c) - I)_{il} = 0$ or, equivalently, $\sum_{i \in S} q_i(c)P(c)_{il} = q_l(c)$. Then summing over all $l \in S$ with $l \neq j$ yields $\sum_{i,l \in S, l \neq j} q_i(c)P(c)_{il} = 1 - q_j(c)$. Furthermore, as $P(c)$ and $q(c)$ are stochastic, we have $\sum_{l \in S} P_{il}(c) = 1$ and $\sum_{i \in S} q_i = 1$ and thus $\sum_{i,l \in S} q_i(c)P_{il}(c) = 1$. Combination of these two observations gives $\sum_{i \in S} q_i(c)P(c)_{ij} = q_j(c)$ and thus $\sum_{i \in S} q_i(c)(P(c) - I)_{ij} = 0$. This proves our claim that we can safely remove one of the $|S|$ constraints of (8.24).

Let $B(c)$ be the $|S| \times |S|$ constraint matrix resulting from the constraints given in (8.24) and (8.25), where we remove the last constraint of (8.24). As indicated, $q(c)$ is the only feasible solution in the polytope defined by $q(c)B(c) = b$ and $q(c) \geq \overline{0}$ with $b = (0, 0, \ldots, 0, 1)$. From the theory of linear programming it now follows that $q(c)$ is a basic feasible solution. Since the neighborhood graph is strongly connected and since $P_{ij}(c) > 0$ for all $i, j \in S$ with $j \in N(i)$, we furthermore have that $q_i(c) > 0$ for all $i \in S$. This implies that each variable $q_i(c)$ is a basic variable

in the basic feasible solution $q(c)$. In other words, $B(c)$ is an invertible matrix and $q(c) = bB^{-1}(c)$. Cramer's rule now states that $q_i(c) = \frac{\det B_i(c)}{\det B(c)}$, where $B_i(c)$ denotes the matrix $B(c)$ with the ith row replaced by the vector b. Note that this is a general expression that holds for any $c > 0$.

Let $Q_i(c) = \det B_i(c)$ and $Q(c) = \det B(c)$. Then $q_i(c)$ can be written as $\frac{Q_i(c)}{Q(c)}$. By definition, the determinant of a finite $n \times n$ matrix M is given by

$$\det M = \sum_{\pi} \mathrm{sgn}(\pi) \prod_{i=1}^{n} M_{i,\pi(i)},$$

where π ranges over all permutations of the columns of M and $\mathrm{sgn}(\pi)$ is a function that assigns either the value -1 or 1 to permutation π. From this it follows that Q_i and Q are polynomials of finite degree in the entries of $B(c)$ and b. By the definition of $P(c)$, we have that the entries of $B(c)$ are linear functions in $\exp\left(-\frac{x_1}{c}\right), \exp\left(-\frac{x_2}{c}\right), \ldots, \exp\left(-\frac{x_L}{c}\right)$ for a finite L, where x_i denotes a cost difference between two neighboring solutions. Combination of these observations yields that Q_i and Q are polynomials of finite degree in $\exp\left(-\frac{x_1}{c}\right), \exp\left(-\frac{x_2}{c}\right), \ldots, \exp\left(-\frac{x_L}{c}\right)$. As a result, the derivatives $Q_i'(c)$ and $Q'(c)$ of $Q_i(c)$ and $Q(c)$ are polynomials of finite degree in $\exp\left(-\frac{x_1}{c}\right), \exp\left(-\frac{x_2}{c}\right), \ldots, \exp\left(-\frac{x_L}{c}\right)$ and their derivatives $\frac{x_1}{c^2}\exp\left(-\frac{x_1}{c}\right), \frac{x_2}{c^2}\exp\left(-\frac{x_2}{c}\right), \ldots, \frac{x_L}{c^2}\exp\left(-\frac{x_L}{c}\right)$. Furthermore, the quotient rule for derivatives gives

$$q_i'(c) = \frac{Q_i'(c)Q(c) - Q_i(c)Q'(c)}{Q^2(c)}. \tag{8.27}$$

The existence of $q(c)$ implies that $Q(c) \neq 0$, which yields that the denominator is positive for all $c \in (0, \infty)$. It now follows that the function $q'(c)$ is continuous on the interval $(0, \infty)$.

We use this result to prove (8.23). First of all, we show that a $\hat{c} > 0$ exists, such that the sign of $q_i'(c)$ is constant for all $c \leq \hat{c}$. This is obviously true if function $q_i'(c)$ is constant. Hence, assume that this is not the case. As observed, the denominator in the definition of $q_i'(c)$ is positive. Hence, the sign of q_i' is determined by the sign of the numerator in (8.27). From the discussion above, it follows that an integer n exists, such that the numerator can be written as

$$\sum_{i=1}^{n} p_i(c^{-1}) \exp(-\lambda_i c^{-1}), \tag{8.28}$$

where $p_i(c^{-1}) = \sum_{j=0}^{m_i} a_{ij} c^{-j}$ with $a_{im_i} \neq 0$ is a polynomial in c^{-1} and λ_i is a positive real number with $1 \leq i \leq n$. For a sufficiently small c, (8.28) is dominated by $a_{im_i} c^{-m_i} \exp(-\lambda_i c^{-1})$, where i is chosen such that λ_i is minimal. In other words, for a sufficiently small c, the sign of (8.28) is equal to the sign of a_{im_i}. As $\{c_k\}_{k=1}^{\infty}$ is non-increasing and as $q_i(c)$ is bounded between 0 and 1, this implies that a $\hat{c} > 0$ exists, such that

$$\sum_{\substack{k=1,2,\ldots \\ c_k \leq \hat{c}}} |q_i(c_k) - q_i(c_{k+1})| \leq 1.$$

Hence, to prove (8.23), it now suffices to show that

$$\sum_{\substack{k=1,2,\dots \\ c_k \geq \hat{c}}} |q_i(c_k) - q_i(c_{k+1})|$$

is finite. Let $c_k, c_{k+1} \geq \hat{c}$. As q_i is continuously differentiable on $(0, \infty)$, we have that $|q_i(c_k) - q_i(c_{k+1})| = |\int_{c_k}^{c_{k+1}} q_i'(c) \, dc|$. Hence, we get

$$\sum_{\substack{k=1,2,\dots \\ c_k \geq \hat{c}}} |q_i(c_k) - q_i(c_{k+1})| \leq \sum_{\substack{k=1,2,\dots \\ c_k \geq \hat{c}}} \int_{c_k}^{c_{k+1}} |q_i'(c)| \, dc.$$

Because $\{c_k\}_{k=1}^{\infty}$ is non-increasing, the right-hand side is at most

$$\int_{\hat{c}}^{c_1} |q_i'(c)| \, dc,$$

which is finite as the continuity of $q_i'(c)$ on $[\hat{c}, c_1]$ implies that $|q_i'(c)|$ is bounded. This concludes the proof of the theorem. □

From this theorem and Corollary 8.2, the following result follows.

Corollary 8.3. *Let (S, f) be an instance of a combinatorial optimization problem with $S^* \neq S$ and S finite. Furthermore, let N be a neighborhood function defined on S that induces a strongly connected, symmetric neighborhood graph with diameter d.*

If a cooling schedule is assumed in which the sequence $\{c_k\}_{k=1}^{\infty}$ of control parameter values is non-increasing and satisfies both $\lim_{k \to \infty} c_k = 0$ and

$$c_k \geq \frac{d\Delta}{\log k}$$

with $\Delta = \max_{i \in S, j \in N(i)} (f(j) - f(i))$, then the inhomogeneous Markov chain associated with a run of simulated annealing is strongly ergodic and the stochastic vector q to which its probability distribution converges satisfies $q_i = 0$ for any non-optimal solution $i \in S$. □

8.6 Non-Convergence Result for Tabu Search

In this chapter we have proved Corollary 8.3, which states that if we have a finite, symmetric, and strongly connected neighborhood graph, then simulated annealing converges to the set of globally optimal solutions whenever the control parameter is lowered slowly enough. In Chapter 7 we presented tabu search as an alternative to simulated annealing for performing a long single walk in the neighborhood graph. One may wonder whether the above-mentioned convergence result also holds for this metaheuristic. As tabu search never stays in the same solution, it obviously does not converge in the sense of simulated annealing, which, after an infinite number of iterations, ends up in an optimal solution. This observation does not, however, answer the question as tabu search stores the best solution encountered. Figure 8.7 nevertheless shows that storing the best solution encountered does not help us. Suppose that tabu search starts in solution 1 and that the tabu list can contain $m \leq n - 4$

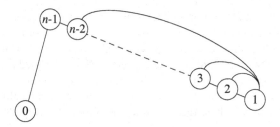

Figure 8.7. Neighborhood graph with n solutions for which tabu search need not converge. The cost of solution i is i for $1 \leq i \leq n - 1$ and 0 for $i = n$.

solutions. The algorithm then cycles through the solutions $1, 2, \ldots, m$ and never visits optimal solution n.

8.7 Bibliographical Notes

In this chapter we make extensive use of Markov chain theory. For more details on this theory we refer you to Feller [1950], Isaacson & Madsen [1976], and Seneta [1981]. In Section 8.3 we first proved Corollary 8.1, which states for the homogeneous model that simulated annealing converges to the set of optimal solutions in the case that the neighborhood graph is strongly connected and the generation matrix satisfies $G_{ij} = G_{ji}$ for all $i, j \in S$. This proof is based on Aarts & Korst [1989]. Next, we generalized this result in Corollary 8.2 by replacing the constraint on the generation matrix by the constraint that the neighborhood graph has to be symmetric. The proof has been provided by Faigle & Kern [1991], who have also presented Example 8.2. In fact, Faigle & Kern [1991] prove a slightly more general result. They show that instead of the neighborhood graph having to be symmetric, it suffices to require that it is weakly reversible. By definition, this means that for any two solutions i and j it holds that the minimum height of a path from i to j is equal to the minimum height of a path from j to i, where the height of a path is defined in Definition 1.13. Although the proof hardly changes, we did not prove this more general result because if a neighborhood graph is not symmetric, then generally it will also not be weakly reversible.

Suppose that for a given set of conditions we know that simulated annealing converges to the set of optimal solutions and suppose that one of these conditions is that the neighborhood graph has to be strongly connected. Van Laarhoven, Aarts & Lenstra [1992] then prove that instead of being strongly connected, it suffices for the neighborhood graph to be only weakly optimally connected. Although at first this seems to be a relevant generalization, this is not the case, as can be seen as follows. From Example 8.2 it follows that the assumption that the neighborhood graph is only strongly connected is not sufficient to guarantee convergence to the set of optimal solutions. The weakest condition for which it has been proved that adding it guarantees such convergence is the weak reversibility condition mentioned above. Hence, the result of Van Laarhoven, Aarts & Lenstra [1992] implies that simulated

annealing converges to the set of optimal solutions if the neighborhood graph is both weakly optimally connected and symmetric, i.e., if the neighborhood graph consists of several strongly connected components that all contain at least one globally optimal solution. However, in this case the application of simulated annealing simply corresponds to applying it to one strongly connected, symmetric component of the neighborhood graph. Hence, the convergence of simulated annealing to the set of optimal solutions then follows directly from the result of Faigle & Kern [1991] presented above.

We note that Van Laarhoven, Aarts & Lenstra [1992] apply their result incorrectly. Lundy & Mees [1986] 'proved' for the homogeneous model that simulated annealing converges to the set of optimal solutions in the case that the generation probability G_{ij} is given by $1/|N(i)|$ if $j \in N(i)$ and by 0 otherwise. However, as observed by Kolonko [1999], in their proof they assumed that the neighborhood graph is symmetric, yet they do not state this assumption explicitly. This leads Van Laarhoven, Aarts & Lenstra [1992] to conclude from the result of Lundy & Mees [1986] that simulated annealing converges to the set of optimal solutions if the neighborhood graph is weakly optimally connected and a neighbor of solution i is selected uniformly at random from its neighborhood $N(i)$. As a consequence, Van Laarhoven, Aarts & Lenstra [1992] claim that simulated annealing converges to an optimal solution for JOB SHOP SCHEDULING when the neighborhood function presented in Section 3.3 is applied. Kolonko [1999] shows that this conclusion is incorrect.

Several authors have studied the convergence of simulated annealing for alternative acceptance probabilities; see Aarts, Korst & Van Laarhoven [1997]. However, in virtually all practical applications the acceptance probabilities chosen are the same as those assumed in this book.

After the statistical analysis given in Section 8.4, which is based on Aarts & Korst [1989], we discussed the convergence of simulated annealing in the realistic inhomogeneous model in Section 8.5. The proof of the main result stated by Theorem 8.13 is based on Anily & Federgruen [1987a] and Anily & Federgruen [1987b] and Example 8.4 is due to Isaacson & Madsen [1976]. From Theorem 8.13 we derived Corollary 8.3. This result is not the strongest known result. Hajek [1988] proved that the constraint $c_k \geq \frac{d\Delta}{\log k}$ in the cooling schedule may be weakened to $c_k \geq \frac{\Gamma}{\log(k+1)}$, where Γ is the maximum depth over all local optima that are not globally optimal, where the depth of a local optimum is defined in Definition 1.14. It can easily be verified that $\Gamma \leq d\Delta$. Furthermore, Hajek [1988] shows that Γ is minimal in the sense that convergence to an optimal solution is no longer guaranteed if $c_k \geq \frac{\Gamma'}{\log(k+1)}$ is assumed with $\Gamma' < \Gamma$. An advantage of the approach we used for proving convergence of simulated annealing is that we relate convergence in the homogeneous model to convergence in the inhomogeneous model. Besides providing insight, Theorem 8.13 yields that if we are interested in whether simulated annealing converges if a given set of conditions is satisfied, including the condition that

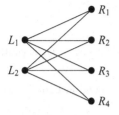

Edge	Weight	Edge	Weight
$\{L_1,R_1\}$	3	$\{L_2,R_1\}$	3
$\{L_1,R_2\}$	4	$\{L_2,R_2\}$	4
$\{L_1,R_3\}$	5	$\{L_2,R_3\}$	6
$\{L_1,R_4\}$	8	$\{L_2,R_4\}$	4

Figure 8.8. Complete bipartite graph with two left nodes and four right nodes, and the weights of its edges.

the neighborhood graph must be strongly connected, then we only have to prove convergence in the homogeneous model.

This chapter focuses on the infinite-time behavior of simulated annealing. Finite-time implementations of simulated annealing are studied by Albrecht [2004], Drost, Jansen & Wegener [2001], Jerrum [1992], Jerrum & Sorkin [1998], Mitra, Romeo & Sangiovanni-Vincentelli [1986], Nolte & Schrader [1996], Nolte & Schrader [1997], Nolte & Schrader [2000], Orosz & Jacobson [2002], Sasaki & Hajek [1988], Sorkin [1991], Steinhöfel, Albrecht & Wong [1998], and Wegener [2005]. Most of them study the finite-time behavior of simulated annealing when applied to a particular combinatorial optimization problem.

As a final remark we note that Exercise 2 is based on Motwani & Raghavan [1995].

8.8 Exercises

1. Consider the complete bipartite graph given in Figure 8.8. This graph consists of two left nodes (L_1 and L_2) and four right nodes (R_1, R_2, R_3, and R_4). A weighted edge exists between each pair of right and left nodes. The weights are also given in Figure 8.8. The goal is to find a maximal matching with minimum weight. A matching is defined as a set of edges in which no pair of edges has a node in common. A matching is maximal if it is not possible to add another edge to the set such that the set remains a matching. Note that in the graph of Figure 8.8 a maximal matching consists of exactly two edges. The weight of a matching is defined as the sum of the weights of all of the edges that the matching contains. As a solution space for this problem, we consider the set of all maximal matchings. Furthermore, we define the following neighborhood. A matching M_1 is a neighbor of matching M_2 (and vice versa) if M_2 can be obtained from M_1 by removing an edge from M_1 and adding a new one.

 a) Draw the neighborhood graph.

 b) Indicate how to change the weight of a single edge, such that the neighborhood function is no longer exact for this example.

We carry out simulated annealing.

 c) Let c be the control parameter and let each neighbor of a matching be generated with equal probability. Give generation matrix G, acceptance matrix A, and the transition matrix P.

 d) Give for each solution the probability that it is found at equilibrium for $c \downarrow 0$, $c = 1$, and $c \to \infty$.

2. Consider random search, which we defined in Section 7.1 as the metaheuristic that performs a random walk through a neighborhood graph.

 a) Model a run of random search as a Markov chain.

We apply random search on a neighborhood graph $G = (V, E)$ that is non-bipartite, finite, symmetric, and strongly connected. Theorem 8.3 states that a finite homogeneous Markov chain is strongly ergodic if it is both irreducible and aperiodic.

 b) Prove the following generalization of Lemma 8.1. An irreducible finite homogeneous Markov chain with transition matrix P is aperiodic if the state transition graph contains a cycle of length l and a cycle of length k with $\gcd(l, k) = 1$, i.e., if a pair i, j of states and a pair k, l of positive integers with $\gcd(l, k) = 1$ exist such that $(P^k)_{ii} > 0$ and $(P^l)_{jj} > 0$.

 c) Prove that the homogeneous Markov chain associated with a run of random search is strongly ergodic.

 d) Prove that the stationary distribution $q(c)$ to which the probability distribution of this Markov chain converges is defined by

$$q_i = \frac{|N(i)|}{2|E|}.$$

We define the *hitting time* h_{ij} as the expected number of steps it takes random search to reach solution j when starting at solution i. In the remainder of this exercise we study the hitting time of solutions for the three types of neighborhood graphs depicted in Figure 8.9. We drop the constraint that the neighborhood graph has to be non-bipartite.

 e) Prove that if the neighborhood graph G is the complete graph on n nodes, then $h_{ij} = n - 1$ for any pair of solutions i, j.

We now consider the neighborhood graph that consists of a single path. We number the solutions from 1 to n in the order they appear on the path. This mean that $N(1) = \{2\}$, $N(n) = \{n-1\}$, and $N(i) = \{i-1, i+1\}$ for $1 < i < n$.

 f) Prove that

$$h_{i,n} - h_{i+1,n} = \begin{cases} 1 & \text{if } i = 1, \\ 2 + h_{i-1,n} - h_{i,n} & \text{if } 1 \leq i < n. \end{cases}$$

 g) Give a closed expression for the difference $h_{i,n} - h_{i+1,n}$ with $1 \leq i < n$.

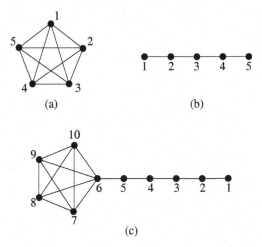

Figure 8.9. (a) Complete neighborhood graph. (b) Neighborhood graph consisting of a single path. (c) Lollipop neighborhood graph.

h) Use (g) to prove $h_{1,n} = (n-1)^2$.

The third type of neighborhood graph that we consider is a combination of the previous two types. It consists of a clique on $\lfloor n/2 \rfloor$ nodes. One of the nodes of this clique is the first node of a path on the remaining nodes; see Figure 8.9(c). We define u and v as the first and last node of the path, where the path is connected to the clique via v. In Figure 8.9(c) we have $u = 1$ and $v = 6$.

i) Prove that $h_{uv} = \Theta(n^2)$ and $h_{vu} = \Theta(n^3)$.

3. In Definition 8.14 we give two equivalent definitions of the ergodic coefficient $\alpha(P)$ of a stochastic $n \times n$ matrix P. In this exercise we prove that the two definitions are indeed equivalent.

a) Prove that for any $1 \le i \le j \le n$

$$\frac{1}{2} \sum_{l=1}^{n} |P_{il} - P_{jl}| = \sum_{l=1}^{n} [P_{il} - P_{jl}]^+,$$

where $[x - y]^+ = \max(0, x - y)$.

b) Prove that

$$\min_{1 \le i \le j \le n} \sum_{l=1}^{n} \min(P_{il}, P_{jl}) = 1 - \frac{1}{2} \max_{1 \le i \le j \le n} \sum_{l=1}^{n} |P_{il} - P_{jl}|.$$

by proving that for any $1 \le i \le j \le n$

$$\sum_{l=1}^{n} [P_{il} - P_{jl}]^+ = 1 - \sum_{l=1}^{n} \min(P_{il}, P_{jl}).$$

A

Graph Theory

In this book we assume that the reader is familiar with graph theory. However, as a reminder and to avoid confusion that might be caused by the fact that different definitions may be used in the literature for the same terminology, we present some basic definitions from graph theory in this appendix. We restrict ourselves to the terminology that is used elsewhere in this book and we do not aim to give an introduction to graph theory. For such an introduction, we refer you to Wilson & Watkins [1990] and Harary [1969].

A graph $G = (V, E)$ consists of a set V of nodes, also called vertices, and a set E containing multisets of V of cardinality two. The elements of E are called edges. An edge $\{i, j\}$ connects nodes i and j. If weights are assigned to the edges and/or the nodes of a graph, then it is called a weighted graph. A graph is said to be planar if it can be drawn in a plane, such that no two edges meet each other except at a node to which they are both connected.

The edge set E may contain loops, which are edges from a node to itself. If we want to emphasize that a graph does not contain loops, we call it a simple graph. If a graph may contain both loops and multiple edges between the same pair of nodes it is called a multigraph. Figure A.1(a) depicts a planar, unweighted graph $G = (V, E)$ with $V = \{1, 2, 3\}$ and $E = \{\{1, 1\}, \{1, 2\}, \{1, 3\}, \{2, 3\}\}$. The edge $\{1, 1\}$ is a loop.

A bipartite graph $G = (V, E)$ is a special kind of graph in which the node set V can be partitioned into two non-empty subsets, such that any edge has one end-point in each of the two subsets. Figure A.1(b) shows a node-weighted bipartite graph, where the weight of a node is written inside the circle by which it is represented.

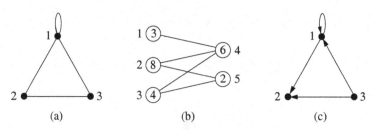

Figure A.1. (a) Graph. (b) Weighted bipartite graph. (c) Digraph.

If the edges of a graph are directed, it is called a directed graph or digraph. Formally, a digraph $D = (V,A)$ is defined by a set V of nodes and a set A of arcs, which are ordered pairs of nodes. An arc (i,j) is directed from node i to node j. Analogously to undirected graphs, we use the term 'simple digraph' to indicate that it may not contain loops. Figure A.1(c) gives a digraph obtained from Figure A.1(a) by orienting the edges. More precisely, the digraph is defined by $V = \{1,2,3\}$ and $E = \{(1,1),(1,2),(3,1),(3,2)\}$.

For an edge $e = \{i,j\}$ in an undirected graph $G = (V,E)$, we say that e is incident with i and j and that i is adjacent to j. The number of edges incident with a node i is called the degree of node i. For an arc $a = (i,j)$ in a digraph $D = (V,A)$, we say that a is incident from i and incident to j. Furthermore, i and j are again called adjacent. The number of arcs incident from a node i is called the outdegree of node i, while the number of arcs incident to node i is called its indegree. For instance, node 3 in the digraph of Figure A.1(c) has outdegree two and indegree zero while the opposite is true for node 2.

Let $G = (V,E)$ be a graph. A walk of length k is a sequence (v_1, v_2, \ldots, v_k) of k nodes, such that $\{v_i, v_{i+1}\} \in E$ for all i with $1 \le i < k$. If we have $v_1 = v_k$ and $k > 1$, then we say that the walk is closed. Hence, the sequence $(1,2,4,2,3)$ defines a walk in the graph of Figure A.2 of length five and $(1,2,4,2,3,1)$ defines a closed walk in this graph of length six. A walk in which all nodes are distinct is called a path, and a closed walk in which all nodes are distinct except for the first and last ones is called a cycle. If a path visits each node in G exactly once, then it is said to be a Hamiltonian path. Analogously, a Hamiltonian cycle of G starts and ends in the same node and visits each other node in G exactly once. It follows that $(1,2,4,6,5,3,1)$ is a Hamiltonian cycle in the graph of Figure A.2. The definitions given easily translate to digraphs.

If graph $G' = (V',E')$ can be obtained from graph $G = (V,E)$ by removing some nodes and/or edges, i.e., if we have $V' \subseteq V$ and $E' \subseteq E$, then we say that G' is a subgraph of G. For instance, the graph of Figure A.1(a) is a subgraph of the graph of Figure A.2. Similarly, we can define a subgraph of a digraph. A graph is said to be connected if it contains a path between any two nodes. If a digraph contains for any pair of nodes i,j a path from i to j and a path from j to i, then it is said to be strongly connected. If only the underlying graph of a digraph D is connected, then we say that D is weakly connected, where the underlying graph of D

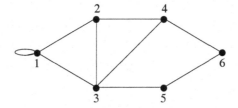

Figure A.2. A graph with exactly two Hamiltonian cycles.

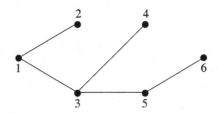

Figure A.3. A spanning tree of the graph of Figure A.2.

is obtained by replacing all its arcs by edges. We define a component of a (di)graph as a non-extendable (weakly) connected sub(di)graph. Note that a (di)graph has a unique subdivision into components. All example graphs given in this appendix are connected. The digraph of Figure A.1(c), however, is only weakly connected as it contains no path from nodes 1 and 2 to node 3 or from node 2 to node 1.

A tree is a connected graph $T = (V, E)$ that does not contain any cycles. If tree T is a subgraph of a graph G that contains all the nodes of G, then it is called a spanning tree of G. Figure A.3 shows a spanning tree of the graph of Figure A.2. Finally, we define an independent set of a graph G as a subset V' of its nodes, such that no two nodes in V' are adjacent.

B

Complexity Theory
and Approximation Algorithms

We present a brief discussion on complexity theory and approximation algorithms. For a more elaborate presentation, we refer to Bovet & Crescenzi [1994], Garey & Johnson [1979], and Papadimitriou [1994].

An important performance criterion of an algorithm is the time it takes to produce a final answer. This time is conveniently expressed as a function of the input size, i.e., the length of the encoding of a problem instance, as we expect the difficulty of a problem to increase with its size. Furthermore, we are especially interested in the worst-case time requirements. Two important reasons for this are that a worst-case time bound gives us a guarantee on when the algorithm will be finished and that the time required on average by an algorithm is generally not substantially better than the time required in the worst case. Formally, this means that the time requirements of an algorithm are specified by a time complexity function f, which for any positive integer n gives the maximum time $f(n)$ that is spent by the algorithm on problem instances of size n. The Turing machine is used as the computational model.

An important distinction is made between algorithms that have a time complexity function that can be bounded from above by a polynomial and algorithms for which the time complexity function cannot be bounded from above by such a polynomial. An algorithm is called a polynomial-time algorithm if it is of the former type and it is called an exponential-time algorithm if it is of the latter type.

A pseudo-polynomial time algorithm is a special case of an exponential-time algorithm: its time complexity function can be bounded from above by a polynomial in both the input size and the magnitude of the largest number that occurs in the problem instance.

It is commonly accepted that polynomial-time algorithms are considered to be efficient, while exponential-time algorithms are not. One reason for this is that for exponential-time algorithms there is an 'explosive' increase in the computing time as the size of the problem instance increases. A second reason is that if, by advances in technology, the speed of a computer were to increase by a given factor, then this would only result in an additive increase in the size of the problem instances that can be handled in a given amount of time for an exponential-time algorithm, whereas for polynomial-time algorithms it would result in a multiplicative increase.

We say that an algorithm solves a combinatorial optimization problem if, on any input x, it either outputs an optimal solution or concludes that x does not define a valid problem instance. In accordance with the distinction between polynomial-time and exponential-time algorithms already discussed, a combinatorial optimization problem is said to be easy if a polynomial-time algorithm exists that solves the problem and it is said to be hard if such an algorithm is not known and it is also unlikely that it exists.

We now discuss the theory of NP-completeness, which formalizes the difference between easy and hard problems. The theory deals with decision problems instead of optimization problems. Decision problems are problems that have the answer either 'yes' or 'no'. Below we indicate that this causes no substantial loss of generality. To relate the computational complexity, i.e., the hardness, of two decision problems, the concept of reducibility is useful. A decision problem Π is said to be polynomially reducible to a problem Π' if a polynomial-time algorithm \mathcal{A} exists that transforms an arbitrary input x of Π into an input $\mathcal{A}(x)$ of Π', such that x defines a yes-instance of Π if and only if $\mathcal{A}(x)$ defines a yes-instance of Π' (if x does not define a valid problem instance, then it is considered as a no-instance). This means that if Π is polynomially reducible to Π', then the existence of a polynomial-time algorithm for Π' implies the existence of a polynomial-time algorithm for Π. In other words, problem Π' is at least as hard as problem Π.

We define P as the class of all easy decision problems, i.e., the problems that can be solved by a polynomial-time algorithm. Furthermore, we define NP as the class of decision problems that can be solved in polynomial time by a non-deterministic computer or, equivalently, for which it holds that each yes-instance has a certificate that can be checked in polynomial time for validity, where the size of the certificate is polynomially bounded in the size of the instance. Note that $P \subseteq NP$.

A problem Π from NP is said to be NP-complete if each problem from NP is polynomially reducible to Π. This means that the NP-complete problems are the hardest problems in NP; once we have an efficient algorithm for any NP-complete problem, we can derive an efficient algorithm for all problems in NP, which implies that $P = NP$. If a problem is NP-complete even when the input is encoded based on the unary notation (in the unary notation the number n is represented by n 1's), then

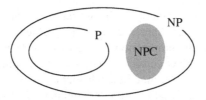

Figure B.1. Relation between the classes P, NP, and NPC in the case that P ≠ NP.

we say the problem is NP-complete in the strong sense. For a strongly NP-complete problem, no pseudo-polynomial time algorithm exists unless P = NP. If an NP-complete problem can be solved by a pseudo-polynomial time algorithm, then we say that it is NP-complete in the ordinary sense. The relation between P, NP, and NPC is visualized in Figure B.1, where NPC is the class of NP-complete decision problems.

SATISFIABILITY was the first problem that was proved to be NP-complete. To show that a problem Π' is NP-complete it now suffices to prove that Π' is in NP and that any known NP-complete problem Π is polynomially reducible to Π'. To prove that Π' is NP-complete in the strong sense it suffices to show that the following three additional requirements hold. (*i*) The NP-complete problem Π from which we reduce is also NP-complete in the strong sense. (*ii*) The size of the instances are not decreased too much by the polynomial-time algorithm \mathcal{A} that reduces Π to Π'. More precisely, this means that the size of a problem instance I from Π is bounded from above by a polynomial in the size of problem instance $\mathcal{A}(I)$ from Π'. (*iii*) If \mathcal{A} increases the maximum value that occurs in an instance, then this increase is bounded from above by a polynomial in the size of the instance from Π. For a slightly less strict proof obligation we refer to Garey & Johnson [1979]. Ausiello et al. [1999] and Garey & Johnson [1979] list a large number of problems that are proved to be NP-complete. Since, until now, no polynomial-time algorithm has been found for any NP-complete problem, it is believed that no such algorithm exists for these problems. In other words, it is believed that P ≠ NP.

We now indicate that restricting ourselves to decision problems does not cause substantial loss of generality. Consider a combinatorial optimization problem that asks for a solution s that satisfies some constraints and that minimizes a given cost function c. With this problem we can associate the decision problem that asks for a given bound B whether a solution s exists that satisfies the same constraints with $c(s) \leq B$. Once we have solved the combinatorial optimization problem, we only need to compare $c(s)$ to B to solve the decision variant. Hence, the combinatorial optimization problem is at least as hard as the decision problem. Consequently, if the decision problem is proved to be NP-complete, then the combinatorial optimization problem is NP-hard, where a problem is called NP-hard if it is at least as hard as any NP-complete problem. Garey & Johnson [1979] discuss that the combinatorial optimization problem is generally also not harder than the associated decision problem. In Lemma B.1 this is proved for one particular problem. Note that although an

efficient algorithm for an NP-hard combinatorial optimization problem is unlikely to exist, it may exist for some special cases of the problem.

We denote the natural optimization counterparts of P and NP by PO and NPO, respectively. The class NPO consists of combinatorial optimization problems for which we can decide in polynomial time whether an input x defines a correct problem instance, for which the cost function is computable in polynomial time, and for which it is decidable in polynomial time whether a string s defines a feasible solution, where the size of each feasible solution is polynomially bounded in the input size. The complexity class PO contains those problems from NPO for which we can derive an optimal solution in polynomial time. For more details we refer to Ausiello et al. [1999] and Bovet and Crescenzi [1994].

As discussed, we can prove the hardness of a problem from NPO by showing that the corresponding decision problem is NP-complete. An interesting question is whether the class of the most difficult problems in NPO coincides with the class of NP-hard problems in NPO. We therefore first need a reduction to relate the complexity of two combinatorial optimization problems from NPO. A possible reduction introduced for this is the Turing reduction. A problem Π is said to be Turing reducible to a problem Π' if a polynomial-time algorithm \mathcal{A} exists that solves Π, where \mathcal{A} may make use of an oracle that solves Π' in constant time. Note that the proposed reduction for decision problems, which is also called a Karp reduction or polynomial transformation, is a special case of a Turing reduction. It corresponds to the case where the oracle is consulted only once, namely at the end of \mathcal{A}. Note that a Turing reduction does indeed have the desired property that if Π is Turing reducible to Π', then Π' is at least as hard as Π.

Using Turing reductions, we can now formalize the notion of an NP-hard optimization problem as a problem to which all decision problems from NP are Turing reducible. Furthermore, we can define the hardest problems in NPO, which are called NPO-complete, as the problems to which all problems in NPO are Turing reducible. Obviously, an NPO-complete problem is also NP-hard. To determine whether an NP-hard problem is NPO-complete, we first analyze the computational complexity of MAXIMUM VARIABLE-WEIGHTED SATISFIABILITY.

Definition B.1 [MAXIMUM VARIABLE-WEIGHTED SATISFIABILITY (MVS)]. Given are a set $U = \{x_1, x_2, \ldots, x_n\}$ of binary variables, a set C of clauses over U, and a weight $w(x_i)$ for each variable x_i. A clause is satisfied by a given truth assignment if and only if at least one of its literals is true. The problem is to find a truth assignment that satisfies all clauses and that maximizes the sum of the weights of the variables that are assigned the value 1, i.e., the value true. □

It can be verified that MVS \in NPO. Let MVS$_{\text{dec}} \in$ NP be the decision problem associated with MVS, i.e., in MVS$_{\text{dec}}$ we are asked whether a feasible truth assignment exists for which the cost is at least a given bound B. The following lemma states that MVS and MVS$_{\text{dec}}$ are equally hard.

Lemma B.1. MVS *and* MVS$_{\text{dec}}$ *are Turing reducible to each other.*

Proof. The claim that MVS$_{\text{dec}}$ is Turing reducible to MVS follows easily from

the observation that MVS_{dec} can be solved by first solving MVS and next checking whether or not the cost of the solution returned is larger than bound B.

We now prove that MVS is Turing reducible to MVS_{dec} as well, i.e., we present a polynomial-time algorithm for MVS that uses an oracle for MVS_{dec}. By using the oracle, we can derive the optimal cost for MVS in polynomial time by performing a binary search on the range $[0, \sum_i w(x_i)]$ of possible outcomes of the cost function to obtain the maximum B for which the decision variant gives a positive answer. We can now construct a polynomial-time algorithm that derives an optimal truth assignment in n steps, where each step contains the algorithm described as a subroutine. In the ith step, the algorithm constructs a problem instance I_0 from the original problem instance by substituting the variables $x_1, x_2, \ldots, x_{i-1}$ with the values we derived for them in the first $i-1$ steps and by substituting x_i with 0. Furthermore, the algorithm defines the problem instance I_1 as I_0, except that x_i is assigned the value 1 instead of the value 0. It next determines the optimal cost for I_0 and I_1 by the subroutine described and, depending on which problem instance implies the largest cost the value 0 or 1 is assigned to x_i. It can easily be verified that the algorithm is optimal and that it runs in polynomial time. As a result, we find that MVS is Turing reducible to MVS_{dec}. \Box

Using Lemma B.1 we can now answer the question stated above.

Theorem B.1. *Let Π be a problem in NPO and let Turing reducibility be used to relate the complexity of problems. Then, Π is NPO-complete if and only if Π is NP-hard.*

Proof. As we have already indicated above, Π is NP-hard if it is NPO-complete. Next, suppose that Π is NP-hard. We show that this implies that it is also NPO-complete. Because Π is NP-hard, we have that MVS_{dec} is Turing reducible to Π. Furthermore, from Lemma B.1 we know that MVS is Turing reducible to MVS_{dec}. Combination of these two observations shows that MVS is Turing reducible to Π. The NPO-completeness of Π now follows from the NPO-completeness of MVS. Ausiello et al. [1999] have proved that MVS is already NPO-complete for an approximation preserving reduction that is stronger than a Turing reduction. \Box

Using the above theorem and lemma, we obtain the following result, which is often stated in the literature, but seldom proved.

Theorem B.2. *PO = NPO if and only if P = NP.*

Proof. If PO = NPO holds, then all NP-hard problems from NPO, such as TSP, can be solved in polynomial time. Hence, PO = NPO implies P = NP. Conversely, if P = NP holds, then the decision problem MVS_{dec} can be solved in polynomial time. According to Lemma B.1 this means that the NP-hard problem MVS can also be solved in polynomial time. Using Theorem B.1, this observation yields that all problems in NPO can be solved in polynomial time, i.e., PO = NPO. \Box

If either the computation time is more important than optimality or an optimal solution is unattainable in a reasonable amount of time because, for example, the

problem is NP-hard (NPO-complete), one can resort to an algorithm that gives a feasible solution fast, but which does not necessarily give an optimal solution. Such an algorithm is called an approximation algorithm. Since for most algorithms the difference between the cost of a solution given by the algorithm and the optimal cost increases with the size of the instance, it is natural to use the relative difference to measure the quality of the solutions given by an approximation algorithm.

Definition B.2. Let \mathcal{A} be an approximation algorithm for a given combinatorial optimization problem. Then R is a *performance bound* of \mathcal{A} for a given problem instance I if

$$\frac{\mathcal{A}(I)}{OPT(I)} \leq R$$

holds in the case of a minimization problem and if

$$\frac{OPT(I)}{\mathcal{A}(I)} \leq R$$

holds in the case of a maximization problem, where $\mathcal{A}(I)$ is the cost of the solution given by \mathcal{A} for instance I and $OPT(I)$ is the optimal cost of I. The smallest possible value of R is called the *performance ratio* of \mathcal{A} for I. □

The terms 'performance bound' and 'performance ratio' are also used to describe the worst-case performance of an approximation algorithm.

Definition B.3. Let \mathcal{A} be an approximation algorithm for a given combinatorial optimization problem. Then R is a performance bound of \mathcal{A} if it is a performance bound of \mathcal{A} for all problem instances. If R is minimal, then R is called the performance ratio of \mathcal{A}. □

Definition B.4. If approximation algorithm \mathcal{A} has performance bound R, then \mathcal{A} is called an *R-approximation algorithm*. □

We already discussed that it is unlikely that the NP-hard problems from NPO can be solved in polynomial time as probably P \neq NP. To indicate the different behaviors of these combinatorial optimization problems with respect to their approximability properties, we identify the following three approximability classes.

- The class APX contains the problems from NPO for which a polynomial-time approximation algorithm exists with some finite constant performance ratio.
- The class PTAS contains the problems from NPO that admit a polynomial-time approximation scheme (PTAS), where a PTAS is an algorithm that takes as input both a problem instance I and a precision $\varepsilon > 0$, and that returns a solution with a performance bound of $1 + \varepsilon$. The running time of a PTAS has to be bounded from above by a polynomial in the size $|I|$ of the problem instance, but may be exponential in $\frac{1}{\varepsilon}$.
- The class FPTAS contains the problems from NPO that admit a fully polynomial-time approximation scheme (FPTAS), where an FPTAS is a PTAS whose running time is bounded from above by a polynomial in both the input size $|I|$ and $\frac{1}{\varepsilon}$.

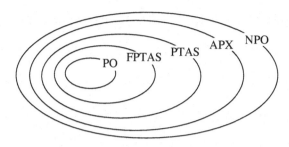

Figure B.2. Relation between the classes PO, FPTAS, PTAS, APX, and NPO in the case that P \neq NP.

Clearly, we have PO \subseteq FPTAS \subseteq PTAS \subseteq APX \subseteq NPO. Furthermore, all inclusions are strict unless P $=$ NP [Ausiello et al., 1999]. The relation between the classes of combinatorial optimization problems presented is visualized in Figure B.2.

Besides the assumption P \neq NP we also prove complexity results in this book that hold under the assumption NP \neq co-NP. The complexity class co-NP is defined as the class NP but with the roles of 'yes' and 'no' exchanged. This means that, while in a decision problem from NP each yes-instance has a certificate that can be checked in polynomial time for validity, a decision problem in co-NP has the property that each no-instance has a disqualification that can be checked in polynomial time for validity. An example of a problem in co-NP is HAMILTONIAN PATH COMPLEMENT, where the problem is to decide whether a graph has *no* Hamiltonian path.

The condition NP \neq co-NP is stronger than the condition P \neq NP, as can be seen as follows. Obviously, if $\Pi \in$ NP \ co-NP, then the complementary problem Π^c is in co-NP \ NP and vice versa. Hence, NP \neq co-NP implies NP $\not\subseteq$ co-NP. Furthermore, as the class P is closed under complementation, we have P \subseteq co-NP. Combination of these two results gives that NP \neq co-NP implies P \neq NP. The opposite, however, need not hold, i.e., it may be the case that NP $=$ co-NP although P \neq NP. This proves that the condition NP \neq co-NP is indeed stronger than the condition P \neq NP. Nevertheless, there is a firm belief that NP \neq co-NP also holds as no one has ever succeeded in showing that HAMILTONIAN PATH COMPLEMENT or any other co-NP complete problem is in NP.

C

PLS-Complete Problems

In this appendix we give a list of local search problems for which it has been proved that they are PLS-complete.

MAX-2SAT

Problem:
 Given are a set U of binary variables and a set C of clauses over U, where the number of literals in each clause is at most two. Furthermore, we are given a weight $w(c) \in \mathbb{N}^+$ for each clause $c \in C$. A clause is satisfied by a truth assignment if and only if at least one of its literals is true. The problem is to find a truth assignment that maximizes the sum of the weights of the clauses that are satisfied.

Solution space:
 The set of all possible truth assignments.

Neighborhood function:
 a) The flip neighborhood function in which truth assignment t' is a neighbor of truth assignment t if and only if t' can be obtained from t by changing the value of one variable.
 b) The Kernighan-Lin neighborhood function, which is obtained by applying variable-depth search to the flip neighborhood function. This means that truth assignment t' is a neighbor of truth assignment t if and only if t' can be obtained from t by performing a sequence of flips, where in each step the

most profitable (least unprofitable) flip is chosen from the flips that change the value of a variable that has not been flipped before. An arbitrary rule may be used for breaking ties.

Result:

MAX-2SAT/flip has been proven to be tightly PLS-complete by Schäffer & Yannakakis [1991] via a tight PLS-reduction from MAX-CUT/flip. In a similar way, we can prove that MAX-2SAT/Kernighan-Lin is tightly PLS-complete via a tight PLS-reduction from MAX-CUT/Kernighan-Lin.

MAX-4SAT WITH BOUNDED VARIABLE OCCURRENCE

Problem:

Given are a set U of binary variables and a set C of clauses over U. The number of literals in each clause is at most four, and each variable occurs at most M times, where M is a constant. Furthermore, we are given a weight $w(c) \in \mathbb{N}^+$ for each clause $c \in C$. A clause is satisfied by a truth assignment if and only if at least one of its literals is true. The problem is to find a truth assignment that maximizes the sum of the weights of the clauses that are satisfied.

Solution space:

The set of all possible truth assignments.

Neighborhood function:

The flip neighborhood function in which truth assignment t' is a neighbor of truth assignment t if and only if t' can be obtained from t by changing the value of one variable.

Result:

Klauck [1996] proves that the induced local search problem is tightly PLS-complete via a tight PLS-reduction from MIN-CIRCUIT/flip. Prior to Klauck, Krentel [1989] proved that the local search problem is tightly PLS-complete in the case that the number of literals per clause is bounded by a constant larger than four.

MAX-CUT

Problem:

Given is an edge-weighted graph $G = (V, E)$ with a weight $w(e) \in \mathbb{N}^+$ for each edge $e \in E$. Find a partition of node set V into two subsets V_1 and V_2, such that the sum of the weights of the edges from E that have one endpoint in V_1 and one endpoint in V_2 is maximal, i.e., such that

$$\sum_{e \in E \cap (V_1 \times V_2)} w(e)$$

is maximal.

Solution space:
The set of all possible partitions of the node set into two subsets.

Neighborhood function:
 a) The flip neighborhood function in which partition (V_1', V_2') is a neighbor of
 partition (V_1, V_2) if and only if (V_1', V_2') can be obtained from (V_1, V_2) by
 moving a single node from one subset to the other one.

 b) The Kernighan-Lin neighborhood function, which is obtained from the flip
 neighborhood function by applying variable-depth search. This means that
 a partition (V_1', V_2') is a Kernighan-Lin neighbor of partition (V_1, V_2) if and
 only if (V_1', V_2') can be obtained from (V_1, V_2) by performing a sequence
 of flips, where in each step the most profitable (least unprofitable) flip is
 chosen from the flips that do not move a node that has already been moved
 before. An arbitrary rule may be used for breaking ties.

Result:
Schäffer & Yannakakis [1991] prove that MAX-CUT/flip is tightly PLS-complete
via a tight PLS-reduction from POS NAE MAX-3SAT/flip. Analogously, one can
derive that MAX-CUT/Kernighan-Lin is also tightly PLS-complete via a tight
PLS-reduction from POS NAE MAX-3SAT/Kernighan-Lin.

For cubic graphs and edge weights that may be both positive and negative, Poljak
[1995] proves that MAX-CUT/flip is solvable in polynomial time. More precisely,
he shows that iterative improvement with the flip neighborhood function runs
in $\mathcal{O}(n^2)$ time, regardless of the pivoting rule used. He also shows that this
time complexity is tight for the pivoting rule that selects an improving move
randomly.

MAXIMUM WEIGHTED INDUCED SUBGRAPH WITH PROPERTY Π

Problem:
Given is a node-weighted graph $G = (V, E)$ with a weight $w(v) \in \mathbb{N}^+$ for each
node $v \in V$. A subset $V' \subseteq V$ of the nodes is said to be feasible if the sub-
graph of G induced by V' has property Π. The property Π must be hereditary,
which means that if any graph satisfies Π, then all node-induced subgraphs of
the graph also satisfy Π, and it must be non-trivial, which means that infinitely
many graphs satisfy Π and infinitely many graphs violate Π. Furthermore, the
property must be verifiable in polynomial time. Find a feasible subset $V' \subseteq V$
that maximizes $\sum_{v \in V'} w(v)$.

Solution space:
The set of all feasible subsets of V that are maximal, i.e., that cannot be extended
without violating property Π.

Neighborhood function:
The extended swap neighborhood function that defines the neighborhood of a

subset V' such that it contains for each $v \in V$ the subset W_v, where W_v is derived from V' by the following algorithm.

If graph $G_0 = (V, \emptyset)$ satisfies Π, then subset W_v is given by V' minus all nodes that are adjacent to v. Otherwise, W_v is given by V' minus all nodes that are not adjacent to v. Next, add v to W_v in the case that $v \notin W_v$ and the subgraph induced by $W_v \cup \{v\}$ satisfies Π. Finally, extend W_v to a maximal feasible subset in a greedy way. This means that we add to W_v the nodes $u \in V \setminus W_v$ for which the subgraph of G induced by $W_v \cup \{u\}$ satisfies Π in a non-increasing weight order.

Result:

For the case that property Π is 'has no edges', it can be shown that the derivation of a local optimum is PLS-complete via a reduction from MAX-2SAT/flip [Schäffer & Yannakakis, 1991]. Note that the subgraph of G induced by a subset V' of the node set has no edges if and only if V' is an independent set of G. Based on this PLS-completeness result, Shimozono [1997] proves PLS-completeness for any Π that is hereditary, non-trivial, and verifiable in polynomial time. For the local search problem to be a member of PLS, it must be possible to compute an initial solution in polynomial time. This is the case because Lewis & Yannakakis [1980] prove that if a property Π is hereditary and non-trivial, then all cliques and/or all independent sets satisfy Π.

METRIC TRAVELING SALESMAN PROBLEM (METRIC TSP)

Problem:

Given are a set $C = \{1, 2, \ldots, n\}$ of n cities and an $n \times n$ matrix d, where entry $d_{ij} \in \mathbb{N}^+$ defines the distance from city i to city j. The matrix d is symmetric and it satisfies the triangle inequality, which means that for all $i, j, k \in C$ we have $d_{ij} = d_{ji}$ and $d_{ij} \leq d_{ik} + d_{kj}$. Find a tour with minimum total tour length, i.e., find a permutation τ of C that minimizes

$$\sum_{i=1}^{n-1} d_{\tau(i), \tau(i+1)} + d_{\tau(n), \tau(1)}.$$

Solution space:

The set of all possible tours.

Neighborhood function:

a) The k-change neighborhood function in which tour τ' is a neighbor of tour τ if and only if τ' can be obtained by replacing at most k edges in τ by new edges.

b) The Lin-Kernighan' neighborhood function, which is a variant of the neighborhood function proposed by Lin & Kernighan [1973]. The Lin-Kernighan' neighborhood $\mathrm{LK}'(\tau)$ of a tour τ is obtained as follows. Let $N_3(\tau)$ be the set of all improving 3-change neighbors of tour τ. If $N_3(\tau) \neq \emptyset$, then $\mathrm{LK}'(\tau)$ is defined as $N_3(\tau)$. Otherwise, $\mathrm{LK}'(\tau)$ is obtained by adding for each pair of adjacent edges $e^{\mathrm{out}}, e^{\mathrm{in}}$ with $e^{\mathrm{out}} \in \tau$, $e^{\mathrm{in}} \notin \tau$,

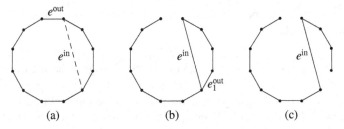

Figure C.1. (a) Tour containing edge e^{out}, but not e^{in}, where the two edges satisfy $d(e^{\text{in}}) < d(e^{\text{out}})$. (b) One-tree obtained by removing edge e^{out} and inserting edge e^{in}. (c) Only Hamiltonian path P that can be obtained from this one-tree by removing an edge adjacent to e^{in}.

and $d(e^{\text{in}}) < d(e^{\text{out}})$ the tour that is constructed by the following algorithm, where $d(e)$ denotes $d_{i,j}$ for edge $e = \{i, j\}$. See Figure C.1(a) for an example of e^{out} and e^{in}.

We first remove edge e^{out} from τ. This results in a Hamiltonian path. We turn this Hamiltonian path into a one-tree (spanning tree plus one edge) by adding edge e^{in}; see Figure C.1(b). We now determine uniquely the edge e_1^{out} adjacent to e^{in}, the removal of which results again in a Hamiltonian path. We remove e_1^{out} from the one-tree and we call the resulting Hamiltonian path P. This operation is visualized in Figure C.1(c).

This ends the initialization phase of our algorithm. The second phase proceeds in iterations, where in iteration i, $i \geq 1$, a Hamiltonian path P_i is constructed from Hamiltonian path P_{i-1} with $P_0 = P$ by adding an edge e_i^{in} and by removing an edge e_{i+1}^{in}. More precisely, iteration i works as follows. We start with the Hamiltonian path P_{i-1}. This path has the property that one of its endpoints is contained in both e^{out} and e^{in}. The other endpoint will change and is correspondingly called the active endpoint. Consider all edges $e_i^{\text{in}} \notin \tau$ leaving the active endpoint, such that the weight of the one-tree obtained by adding e_i^{in} to P_{i-1} is smaller than the length of the original tour τ, i.e., such that

$$\sum_{j=0}^{i} (d(e_j^{\text{out}}) - d(e_j^{\text{in}})) > 0,$$

where $e_0^{\text{out}} = e^{\text{out}}$ and $e_0^{\text{in}} = e^{\text{in}}$. If no such edges exist, then we end phase two of the algorithm and proceed with the last phase. If such edges exist, then we choose from these edges one that maximizes $d(e_{i+1}^{\text{out}}) - d(e_i^{\text{in}})$, where e_{i+1}^{out} is determined uniquely by e_i^{in} in the same way as e_1^{out} is determined uniquely by e^{in}. Ties are broken lexicographically. We add edge e_i^{in} to P_{i-1} and next remove edge e_{i+1}^{out}. We now start the next iteration.

In the third and last phase of the algorithm we determine for each derived Hamiltonian path P_i the length of the tour obtained by connecting the two

endpoints. The best of these tours is returned by the algorithm and thus added to $LK'(\tau)$.

Result:

To prove that METRIC TSP and one of the above neighborhood functions yield a (tightly) PLS-complete local search problem, it suffices to prove this result for SYMMETRIC TSP. The triangle inequality can be satisfied by adding a large value to each distance d_{ij}. Krentel [1989] proves that METRIC TSP/k-change is tightly PLS-complete for a huge value of k. He proves this result via a tight PLS-reduction from the local search problem induced by MAX-4SAT WITH BOUNDED VARIABLE OCCURRENCE and the flip neighborhood function. Via a tight PLS-reduction from MAX-2SAT/flip Papadimitriou [1992] and Papadimitriou, Schäffer & Yannakakis [1990] prove that METRIC TSP/Lin-Kernighan' is also tightly PLS-complete.

MIN-CIRCUIT

Problem:

Given is a Boolean circuit D with n input nodes x_1, x_2, \ldots, x_n and m output nodes y_1, y_2, \ldots, y_m. Find a solution $s \in \{0,1\}^n$ that, when given as an input to D, minimizes the cost function

$$f(s) = \sum_{i=1}^{m} 2^{i-1} y_i.$$

Solution space:

The set $\{0,1\}^n$ of all possible input vectors of the circuit.

Neighborhood function:

The flip neighborhood function in which solution (vector) s' is a neighbor of s if and only if s' can be obtained by changing the value of a single bit in vector s.

Result:

MIN-CIRCUIT/flip is tightly PLS-complete as shown by Johnson, Papadimitriou & Yannakakis [1988] and Yannakakis [1997]. The variant MAX-CIRCUIT/flip, in which the cost function has to be maximized instead of minimized, is also tightly PLS-complete [Johnson, Papadimitriou & Yannakakis, 1988; Yannakakis, 1997]. For a proof of these two results we refer to Theorems 6.4 and 6.10. The generic term for MIN-CIRCUIT/flip and MAX-CIRCUIT/flip is CIRCUIT/flip.

MINIMUM WEIGHTED INDEPENDENT DOMINATING SET FOR GRAPHS WITH BOUNDED DEGREE

Problem:

Given is a node-weighted graph $G = (V, E)$ with a weight $w(v) \in \mathbb{N}^+$ for each node $v \in V$ and with a maximum degree of at most B, where B is a constant. A

subset $V' \subseteq V$ of the nodes is called an independent set if no two nodes in V' are joined by an edge in E, and V' is called a dominating set if each node v in $V \setminus V'$ is joined by an edge in E to at least one node in V'. Find an independent, dominating set $V' \subseteq V$ that minimizes $\sum_{v \in V'} w(v)$.

Solution space:
The set of all independent, dominating sets.

Neighborhood function:
The k-flip neighborhood function in which the neighborhood of an independent, dominating set V' is obtained by adding k_1 nodes to V' and by removing k_2 nodes from V', where k_1 and k_2 are two non-negative integers that satisfy $k_1 + k_2 \leq k$.

Result:
Klauck [1996] proves that the induced local search problem is tightly PLS-complete for a sufficiently large k. He proves this result via a reduction from MAX-4SAT WITH BOUNDED VARIABLE OCCURRENCE and the flip neighborhood function.

POSITIVE NOT-ALL-EQUAL MAX-3SAT (POS NAE MAX-3SAT)

Problem:
Given are a set U of binary variables, a set C of clauses over U, and a weight $w(c) \in \mathbb{N}^+$ for each clause $c \in C$. A clause does not contain negative literals, and the number of positive literals plus the number of constants in each clause is at most three. A clause is satisfied by some truth assignment if and only if it contains at least one literal/constant with value true and at least one literal/constant with value false. The problem is to find a truth assignment that maximizes the sum of the weights of clauses that are satisfied.

Solution space:
The set of all possible truth assignments.

Neighborhood function:
a) The flip neighborhood function in which a truth assignment t' is a neighbor of truth assignment t if and only if t' can be obtained from t by changing the value of one variable.

b) The Kernighan-Lin neighborhood function, which is obtained by applying variable-depth search to the flip neighborhood function. This means that this neighborhood function is defined, such that truth assignment t' is a neighbor of truth assignment t if and only if t' can be derived from t by performing a sequence of flips, where in each step the most profitable (least unprofitable) flip is chosen from the flips that change the value of a variable that has not been flipped before. An arbitrary rule may be used for breaking ties.

Result:
Schäffer & Yannakakis [1991] prove that POS NAE MAX-3SAT/flip is

tightly PLS-complete, and Yannakakis [1997] proves that POS NAE MAX-3SAT/Kernighan-Lin is tightly PLS-complete. Both results are proved via a reduction from CIRCUIT/flip. For a proof of the latter result, we refer to Lemma 6.2 and Theorem 6.10.

PURE NASH EQUILIBRIA IN CONGESTION GAMES

Problem:
Given is a congestion game, which is defined by a set $\{1, 2, \ldots, n\}$ of n players, a set E of resources, and for each player i an action set $S_i \subseteq 2^E$. Furthermore, we are given a delay function $d : E \times \{1, 2, \ldots, n\} \to \mathbb{Z}$, such that $d(e, j)$ is non-decreasing in j. An action combination is an element of $S_1 \times S_2 \times \cdots \times S_n$, and it defines for each player the set of resources it picks. For any action combination $s = (s_1, s_2, \ldots, s_n)$ we define $f_s(e) = |\{i \mid e \in s_i\}|$ as the congestion of resource e and $u_i(s) = -\sum_{e \in s_i} d(e, f_s(e))$ as the payoff function of player i. Find an action combination s for the congestion game that minimizes the cost function

$$\sum_{e \in E} \sum_{j=1}^{f_s(e)} d(e, j).$$

Solution space:
The set of all possible action combinations.

Neighborhood function:
The switch neighborhood function in which action combination s' is a neighbor of action combination s if and only if s' can be obtained from s by changing the action of one player.

Result:
An action combination s is called a pure Nash equilibrium of the congestion game if changing the action of a player i does not improve the payoff of that player, i.e., if for any player i we have $u_i(s) \geq u_i(s_i')$, where s_i' only differs from s in the action of player i. The local optima with respect to the switch neighborhood function correspond to pure Nash equilibria for the given congestion game [Fabrikant, Papadimitriou & Talwar, 2004; Rosenthal, 1973]. Via a tight PLS-reduction from POS NAE MAX-3SAT/flip Fabrikant, Papadimitriou & Talwar [2004] prove that the problem of finding a local optimum is tightly PLS-complete. They also prove that this remains true if the congestion game is symmetric, which means that each player has the same action set.

A special type of congestion game is a network congestion game. In such a game the action sets are represented by paths in a directed graph. To be more precise, we are given a directed graph and, for each player i, a source node v_i and a sink node w_i. The arcs now correspond to resources, and each path from v_i to w_i defines an action that is available to player i.

Fabrikant, Papadimitriou & Talwar [2004] prove that finding a switch local optimum is tightly PLS-complete for asymmetric network congestion games and that it can be done in polynomial time for symmetric network congestion games.

STABLE CONFIGURATION

Problem:

Given is a Hopfield network that is defined by an edge-weighted simple graph $G = (V, E)$ with a weight $w(e) \in \mathbb{Z}$ for each edge $e \in E$ and a threshold value T_v for each node $v \in V$. A configuration $f : V \to \{-1, 1\}$ assigns to each node v the value -1 or 1. The value $f(v)$ is called the state of node v. Find a configuration f that maximizes the cost function

$$\sum_{\{u,v\}\in E} w(\{u,v\})f(u)f(v) - \sum_{v\in V} T_v f(v).$$

Solution space:

The set of all possible configurations.

Neighborhood function:

The flip neighborhood function in which configuration f' is a neighbor of configuration f if and only if f' can be obtained from f by changing the state of exactly one node.

Result:

Via a tight PLS-reduction from MAX-CUT/flip, Schäffer & Yannakakis [1991] prove that the problem of finding a local optimum is already tightly PLS-complete if $T_v = 0$ for each node $v \in V$ and if all weights on the edges are negative. The set of local optima corresponds to the set of stable configurations of the Hopfield network [Hopfield, 1982], where a configuration f is said to be stable if we have

$$f(v) = 1 \Rightarrow \sum_{\{u,v\}\in E} w(\{u,v\})f(u) \geq T_v$$

and

$$f(v) = -1 \Rightarrow \sum_{\{u,v\}\in E} w(\{u,v\})f(u) \leq T_v.$$

For more details we refer to Section 2.6.

UNIFORM GRAPH PARTITIONING (UGP)

Problem:

Given is an edge-weighted graph $G = (V, E)$ with $|V| = 2n$ and a weight $w(e) \in \mathbb{N}$ for each edge $e \in E$. Find a partition of V into two subsets V_1 and V_2 with $|V_1| = |V_2| = n$, such that the sum of the weights of the edges that have one endpoint in V_1 and one endpoint in V_2 is minimal, i.e., such that

$$\sum_{e\in E\cap(V_1\times V_2)} w(e)$$

is minimal.

Solution space:

The set of all possible partitions.

Neighborhood function:

a) The swap neighborhood function in which partition (V_1', V_2') is a neighbor of partition (V_1, V_2) if and only if (V_1', V_2') can be obtained from (V_1, V_2) by interchanging a node from V_1 with a node from V_2.

b) The Kernighan-Lin neighborhood function [Kernighan & Lin, 1970], which is obtained by applying variable-depth search to the swap neighborhood function. This means that partition (V_1', V_2') is a Kernighan-Lin neighbor of partition (V_1, V_2) if and only if (V_1', V_2') can be derived from (V_1, V_2) by performing a sequence of swaps, where in each step the most profitable (least unprofitable) swap is chosen from the swaps that operate on nodes that have not been swapped before. An arbitrary rule may be used for breaking ties.

c) The Fiduccia-Mattheyses neighborhood function [Fiduccia & Mattheyses, 1982], which differs from the Kernighan-Lin neighborhood function in that each swap is performed in two steps. This means that partition (V_1', V_2') is a Fiduccia-Mattheyses neighbor of (V_1, V_2) if and only if (V_1', V_2') can be obtained from (V_1, V_2) by performing a sequence of steps, where a single step proceeds as follows. First of all, we look for a node for which a move to the other subset would be most profitable (least unprofitable), where we only consider nodes that have not been moved before. We move this node to the other subset. Next, we balance the partition by performing the most profitable (least unprofitable) move in the opposite direction where, again, we only consider nodes that have not been moved before. As in the Kernighan-Lin neighborhood function, an arbitrary rule may be used for breaking ties.

d) The FM-swap neighborhood function. This is a simplification of the Fiduccia-Mattheyses neighborhood function: a partition contains only one neighbor, namely the first Fiduccia-Mattheyses neighbor.

Result:

Schäffer & Yannakakis [1991] prove that UGP/swap and UGP/FM-swap are tightly PLS-complete via a tight PLS-reduction from MAX-CUT/flip. In a similar way, MAX-CUT/Kernighan-Lin can be reduced to both UGP/Kernighan-Lin and UGP/Fiduccia-Mattheyses, which proves that these two problems are also tightly PLS-complete. In Theorems 6.5 and 6.10 we prove that UGP/Kernighan-Lin is tightly PLS-complete via a tight PLS-reduction from POS NAE MAX-3SAT/Kernighan-Lin. This proof is based on Yannakakis [1997]. Johnson, Papadimitriou & Yannakakis [1988] were the first to prove that UGP/Kernighan-Lin is PLS-complete.

0-1 INTEGER PROGRAMMING

Problem:

Given are two vectors $b \in \mathbb{Z}^m$ and $c \in \mathbb{N}^n$ and an $m \times n$ matrix A with elements $a_{ij} \in \mathbb{Z}$. A vector $x \in \{0,1\}^n$ is called a feasible solution if it satisfies $Ax \geq b$. Find a feasible solution x that minimizes $c^T x$.

Solution space:
 The set of all feasible solutions.

Neighborhood function:
 The k-flip neighborhood function in which solution (vector) x' is a neighbor of x if and only if x' can be obtained by changing the value of at most k bits in x.

Result:
 Klauck [1996] proves that the induced local search problem is tightly PLS-complete for a sufficiently large k. This result is proved via a reduction from MAX-4SAT WITH BOUNDED VARIABLE OCCURRENCE with the flip neighborhood function.

Bibliography

AARTS, E.H.L., AND J. KORST [1989], *Simulated Annealing and Boltzmann Machines*, Wiley.

AARTS, E.H.L., J. KORST, AND P.J.M. VAN LAARHOVEN [1997], Simulated annealing, in: E.H.L. Aarts and J.K. Lenstra (eds.), *Local Search in Combinatorial Optimization*, Wiley, 91–120.

AARTS, E.H.L., AND J.K. LENSTRA (eds.) [1997], *Local Search in Combinatorial Optimization*, Wiley.

AARTS, E.H.L., AND M. VERHOEVEN [1997], Local search, in: M. Dell'Amico, F. Maffioli, and S. Martello (eds.), *Annotated Bibliographies in Combinatorial Optimization*, Wiley, 163–180.

ALBRECHT, A.A. [2004], A problem-specific convergence bound for simulated annealing-based local search, *Proceedings of the International Conference on Computational Science and its Applications*, Assisi, Italy, 405–414.

ALIMONTI, P. [1995], Non-oblivious local search for graphs and hypergraph coloring problems, *Proceedings of the 21st Workshop on Graph-Theoretic Concepts in Computer Science*, Aachen, Germany, 167–180.

ALIMONTI, P. [1997], Non-oblivious local search for MAX 2-CCSP with application to MAX DICUT, *Proceedings of the 23rd Workshop on Graph-Theoretic Concepts in Computer Science*, Berlin, Germany, 2–14.

ANGEL, E., AND V. ZISSIMOPOULOS [1998], On the quality of local search for the quadratic assignment problem, *Discrete Applied Mathematics* **82**, 15–25.

ANILY, S., AND A. FEDERGRUEN [1987a], Ergodicity in parametric nonstationary Markov chains: An application, *Operations Research* **35**, 867–874.

ANILY, S., AND A. FEDERGRUEN [1987b], Simulated annealing methods with general acceptance probabilities, *Journal of Applied Probability* **24**, 657–667.

APPEL, K., AND W. HAKEN [1977], Every planar map is four colorable. Part I: Discharging, *Illinois Journal of Mathematics* **21**, 429–490.

APPEL, K., AND W. HAKEN [1989], *Every Planar Map is Four Colorable*, American Mathematical Society, Contemporary Mathematics 98.

APPEL, K., W. HAKEN, AND J. KOCH [1977], Every planar map is four colorable. Part II: Reducibility, *Illinois Journal of Mathematics* **21**, 491–567.

ARKIN, E.M., AND R. HASSIN [1998], On local search for weighted k-set packing, *Mathematics of Operations Research* **23**, 640–648.

ARMSTRONG, D.E., AND S.H. JACOBSON [2003], Studying the complexity of global verification for NP-hard discrete optimization problems, *Journal of Global Optimization* **27**, 83–96.

ARORA, S. [1998], Polynomial-time approximation schemes for Euclidean traveling salesman and other geometric problems, *Journal of the ACM* **45**, 753–782.

ARYA, V., N. GARG, R. KHANDEKAR, A. MEYERSON, K. MUNAGALA, AND V. PANDIT

[2004], Local search heuristic for k-median and facility location problems, *SIAM Journal on Computing* **33**, 544–562.

AUSIELLO, G., P. CRESCENZI, G. GAMBOSI, V. KANN, A. MARCHETTI-SPACCAMELA, AND M. PROTASI [1999], *Complexity and Approximation: Combinatorial Optimization Problems and Their Approximability Properties*, Springer.

BAFNA, V., B. NARAYANAN, AND R. RAVI [1996], Nonoverlapping local alignments (weighted independent sets of axis-parallel rectangles), *Discrete Applied Mathematics* **71**, 41–53.

BEALE, E.M.L. [1988], *Introduction to Optimization*, Wiley.

BELLARE, M., O. GOLDREICH, AND M. SUDAN [1998], Free bits, PCPs, and nonapproximability - towards tight results, *SIAM Journal on Computing* **27**, 804–915.

BERMAN, A., AND R.J. PLEMMONS [1979], *Nonnegative matrices in the mathematical sciences*, Academic Press.

BERMAN, P. [2000], A $d/2$ approximation for maximum weight independent set in d-claw free graphs, *Nordic Journal of Computing* **7**, 178–184.

BERN, M., AND P. PLASSMANN [1989], The Steiner problem with edge lengths 1 and 2, *Information Processing Letters* **32**, 171–176.

BOCK, F. [1958], An algorithm for solving 'traveling-salesman' and related network optimization problems: Abstract, presented at the 14^{th} National Meeting of the Operations Research Society of America.

BOVET, D., AND P. CRESCENZI [1994], *Introduction to the Theory of Complexity*, Prentice-Hall.

BOYKOV, Y., O. VEKSLER, AND R. ZABIH [1998], Markov random fields with efficient approximations, *Proceedings of the 1998 IEEE Conference on Computer Vision and Pattern Recognition*, Santa Barbara, CA, 648–655.

BOYKOV, Y., O. VEKSLER, AND R. ZABIH [2001], Fast approximate energy minimization via graph cuts, *IEEE Transactions on Pattern Analysis and Machine Intelligence* **23**, 1222–1239.

BRUCKER, P. [1995], *Scheduling Algorithms*, Springer.

BRUCKER, P., J. HURINK, AND F. WERNER [1997], Improving local search heuristics for some scheduling problems, *Discrete Applied Mathematics* **72**, 47–69.

BRÜGGEMANN, T., J. MONNOT, AND G.J. WOEGINGER [2003], Local search for the minimum label spanning tree problem with bounded color classes, *Operations Research Letters* **31**, 195–201.

BYLKA, S., A. IDZIK, AND Z. TUZA [1999], Maximum cuts: Improvements and local algorithmic analogues of the Edwards-Erdös inequality, *Discrete Mathematics* **194**, 39–58.

CERNY, V. [1985], Thermodynamical approach to the traveling salesman problem: An efficient simulation algorithm, *Journal of Optimization Theory and Applications* **45**, 41–51.

CHANDRA, B., AND M.M. HALLDÓRSSON [2001], Greedy local improvement and weighted packing approximation, *Journal of Algorithms* **39**, 223–240.

CHANDRA, B., H. KARLOFF, AND C. TOVEY [1999], New results on the old k-opt algorithm for the traveling salesman problem, *SIAM Journal on Computing* **28**, 1998–2029.

CHARIKAR, M., AND S. GUHA [1999], Improved combinatorial algorithms for the facility location and k-median problems, *Proceedings of the 40^{th} Annual Symposium on Foundations of Computer Science*, New York City, NY, 378–388.

CHIBA, N., T. NISHIZEKI, AND N. SAITO [1981], A linear 5-coloring algorithm of planar graphs, *Journal of Algorithms* **2**, 317–327.

CHO, Y., AND S. SAHNI [1980], Bounds for list schedules on uniform processors, *SIAM Journal on Computing* **9**, 91–103.

CHRISTOFIDES, N. [1976], *Worst-case analysis of a new heuristic for the traveling salesman problem*, Technical Report 388, Graduate School of Industrial Administration, Carnegie-Mellon University, Pittsburgh, PA.

COOK, W.J., W.H. CUNNINGHAM, W.R. PULLEYBLANK, AND A. SCHRIJVER (eds.) [1998], *Combinatorial Optimization*, Wiley.

CORNE, D., M. DORIGO, AND F. GLOVER (eds.) [1999], *New Ideas in Optimization*, McGraw-Hill.

CORNUEJOLS, G., AND G.L. NEMHAUSER [1978], Tight bounds for Christofides' traveling salesman heuristic, *Mathematical Programming* **14**, 116–121.

CROES, G.A. [1958], A method for solving traveling-salesman problems, *Operations Research* **6**, 791–812.

DANTSIN, E., A. GOERDT, E.A. HIRSCH, R. KANNAN, J. KLEINBERG, C.H. PAPADIMITRIOU, P. RAGHAVAN, AND U. SCHÖNING [2002], A deterministic $(2 - \frac{2}{k+1})^n$ algorithm for k-SAT based on local search, *Theoretical Computer Science*, 69–83.

DAVIS, J.S., AND J.J. KANET [1993], Single-machine scheduling with early and tardy completion costs, *Naval Research Logistics* **40**, 85–101.

DEINEKO, V.G., M. HOFFMANN, Y. OKAMOTO, AND G.J. WOEGINGER [2004], The traveling salesman problem with few inner points, *Proceedings of the 10^{th} International Computing and Combinatorics Conference*, Jeju Island, Korea, 268–277.

DELL'AMICO, M., F. MAFFIOLI, AND S. MARTELLO (eds.) [1997], *Annotated Bibliographies in Combinatorial Optimization*, Wiley.

DELL'AMICO, M., AND M. TRUBIAN [1993], Applying tabu search to the job-shop scheduling problem, *Annals of Operations Research* **41**, 231–252.

DIJKSTRA, E.W. [1959], A note on two problems in connexion with graphs, *Numerische Mathematik* **1**, 269–271.

DROST, S., T. JANSEN, AND I. WEGENER [2001], Dynamic parameter control in simple evolutionary algorithms, *Proceedings of the 6^{th} Workshop on Foundations of Genetic Algorithms*, Charlottesville, VA, 275–294.

DUECK, G., AND T. SCHEUER [1990], Threshold accepting: A general purpose optimization algorithm appearing superior to simulated annealing, *Journal of Computational physics* **90**, 161–175.

FABRIKANT, A., C.H. PAPADIMITRIOU, AND K. TALWAR [2004], The complexity of pure Nash equilibria, *Proceedings of the 36^{th} ACM Symposium on Theory of Computing*, Chicago, IL, 604–612.

FAIGLE, U., AND W. KERN [1991], Note on the convergence of simulated annealing algorithms, *SIAM Journal on Control and Optimization* **29**, 153–159.

FEIGE, U., AND J. KILIAN [1998], Zero knowledge and the chromatic number, *Journal of Computer and System Sciences* **57**, 187–199.

FEIGLE, U., M. KARPINSKI, AND M. LANGBERG [2002], Improved approximation of MAX-CUT on graphs of bounded degree, *Journal of Algorithms* **43**, 201–219.

FELLER, W. [1950], *An Introduction to Probability Theory and Its Applications; Volume 1*, Wiley.

FEO, T.A., AND M.G.C. RESENDE [1989], A probabilistic heuristic for a computationally difficult set covering problem, *Operations Research Letters* **8**, 67–71.

FESTA, P., AND M.G.C. RESENDE [2002], GRASP: An annotated bibliography, in: C.C. Ribeiro and P. Hansen (eds.), *Essays and Surveys on Metaheuristics*, Kluwer Academic Publishers, 325–367.

FIDUCCIA, C.M., AND R.M. MATTHEYSES [1982], A linear-time heuristic for improving network partitions, *Proceedings of the 19th ACM/IEEE Design Automation Conference*, Los Alamitos, CA, 175–181.

FINN, G., AND E. HOROWITZ [1979], A linear-time approximation algorithm for multiprocessor scheduling, *BIT* **19**, 312–320.

FISCHER, S.T. [1995], A note on the complexity of local search problems, *Information Processing Letters* **53**, 69–75.

FLOOD, M.M. [1956], The traveling salesman problem, *Operations Research* **4**, 61–75.

FOULDS, L.R. [1984], *Combinatorial Optimization for Undergraduates*, Springer.

GAREY, M.R., R.L. GRAHAM, AND D.S. JOHNSON [1976], Some NP-complete geometric problems, *Proceedings of the 8th ACM Symposium on Theory of Computing*, New York, NY, 10–22.

GAREY, M.R., R.L. GRAHAM, AND D.S. JOHNSON [1977], The complexity of computing Steiner minimal trees, *SIAM Journal on Applied Mathematics* **32**, 835–859.

GAREY, M.R., AND D.S. JOHNSON [1979], *Computers and Intractability: A Guide to the Theory of NP-Completeness*, W.H. Freeman and Company.

GAREY, M.R., D.S. JOHNSON, AND R. SETHI [1976], The complexity of flowshop and jobshop scheduling, *Mathematics of Operations Research* **1**, 117–129.

GAREY, M.R., D.S. JOHNSON, AND L. STOCKMEYER [1976], Some simplified NP-complete graph problems, *Theoretical Computer Science* **1**, 237–267.

GAREY, M.R., R.E. TARJAN, AND G.T. WILFONG [1988], One-processor scheduling with symmetric earliness and tardiness penalties, *Mathematics of Operations Research* **13**, 330–348.

GIBBONS, A. [1985], *Algorithmic Graph Theory*, Cambridge University Press.

GLOVER, F. [1986], Future paths for integer programming and links to artificial intelligence, *Computers and Operations Research* **13**, 533–549.

GLOVER, F., AND G. KOCHENBERGER (eds.) [2003], *Handbook of Metaheuristics*, Kluwer Academic Publishers.

GLOVER, F., AND M. LAGUNA [1997], *Tabu Search*, Kluwer Academic Publishers.

GOLDBERG, D.E. [1989], *Genetic Algorithms in Search, Optimization, and Machine Learning*, Addison-Wesley.

GONZALEZ, T., AND S. SAHNI [1978], Flowshop and jobshop schedules: Complexity and approximation, *Operations Research* **26**, 36–52.

GRAHAM, R.L., E.L. LAWLER, J.K. LENSTRA, AND A.H.G. RINNOOY KAN [1979], Optimization and approximation in deterministic sequencing and scheduling: A survey, *Annals of Discrete Mathematics* **5**, 287–326.

GRIMMETT, G.R., AND D.R. STIRZAKER [1992], *Probability and Random Processes*, Oxford University Press.

GUPTA, A., AND E. TARDOS [2000], A constant factor approximation algorithm for a class of classification problems, *Proceedings of the 32nd ACM Symposium on Theory of Computing*, Portland, OR, 652–658.

GUTIN, G., AND A.P. PUNNEN (eds.) [2002], *The Traveling Salesman Problem and Its Variations*, Kluwer Academic Publishers.

HAJEK, B. [1988], Cooling schedules for optimal annealing, *Mathematics of Operations Research* **13**, 311–329.

HAKEN, A. [1989], Connectionist networks that need exponential time to stabilize, unpublished manuscript, Department of Computer Science, University of Toronto, Toronto, Canada.

HALLDÓRSSON, M.M. [1993], A still better performance guarantee for approximate graph coloring, *Information Processing Letters* **45**, 19–23.

HALLDÓRSSON, M.M. [1995], A still better performance guarantee for approximate graph coloring, *Proceedings of the 6th ACM-SIAM Symposium on Discrete Algorithms*, San Francisco, CA, 160–169.

HALLDÓRSSON, M.M., AND H.C LAU [1997], Low-degree graph partitioning via local search with applications to constraint satisfaction, max cut, and 3-coloring, *Journal of Graph Algorithms and Applications* **1**, 1–13.

HANSEN, P. [1986], Presented at the Congress on Numerical Methods in Combinatorial Optimization, Capri, Italy.

HANSEN, P., AND B. JAUMARD [1990], Algorithms for the maximum satisfiability problem, *Computing* **44**, 279–303.

HARARY, F. [1969], *Graph Theory*, Addison-Wesley.

HASSIN, R., A. LEVIN, AND D. MORAD [2003], Lexicographic local search and the p-center problem, *European Journal of Operational Research* **151**, 265–279.

HEAWOOD, P.J. [1890], Map colour theorem, *Quarterly Journal of Pure and Applied Mathematics* **24**, 332–338.

HIRSCH, E.A. [2003], Worst-case study of local search for MAX-k-SAT, *Discrete Applied Mathematics* **130**, 173–184.

HOLLAND, J.H. [1975], *Adaptation in Natural and Artificial Systems*, University of Michigan Press.

HOOS, H.H., AND T. STÜTZLE [2004], *Stochastic Local Search: Foundations and Applications*, Morgan Kaufmann.

HOPFIELD, J.J. [1982], Neural networks and physical systems with emergent collective computational abilities, *Proceedings of the National Academy of Sciences of the United States of America* **79**, 2554–2558.

HURKENS, C.A.J., AND A. SCHRIJVER [1989], On the size of systems of sets every t of which have an SDR, with an application to the worst-case ratio of heuristics for packing problems, *SIAM Journal on Discrete Mathematics* **2**, 68–72.

HURKENS, C.A.J., AND T. VREDEVELD [2003], Local search for multiprocessor scheduling: How many moves does it take to a local optimum?, *Operations Research Letters* **31**, 137–141.

ISAACSON, D., AND R. MADSEN [1976], *Markov Chains*, Wiley.

JERRUM, M. [1992], Large cliques elude the Metropolis process, *Random Structures and Algorithms* **3**, 347–359.

JERRUM, M., AND G.B. SORKIN [1998], The Metropolis algorithm for graph bisection, *Discrete Applied Mathematics* **82**, 155–175.

JOHNSON, D.S. [1990], Local optimization and the traveling salesman problem, *Proceedings of the 17th Colloquium on Automata, Languages, and Programming*, Warwick University, England, 446–461.

JOHNSON, D.S., AND L.A. MCGEOCH [1997], The traveling salesman problem: A case study, in: E.H.L. Aarts and J.K. Lenstra (eds.), *Local Search in Combinatorial Optimization*, Wiley, 215–310.

JOHNSON, D.S., C.H. PAPADIMITRIOU, AND M. YANNAKAKIS [1988], How easy is local search?, *Journal of Computer and System Sciences* **37**, 79–100.

Bibliography

KANUNGO, T., D.M. MOUNT, N.S. NETANYAHU, C.D. PIATKO, R. SILVERMAN, AND A.Y. WU [2002], A local search approximation algorithm for k-means clustering, *Proceedings of the 18th Annual Symposium on Computational Geometry*, Barcelona, Spain, 10–18.

KEMPE, A.B. [1879], On the geographical problem of the four colors, *American Journal of Mathematics* 2, 193–200.

KERNIGHAN, B.W., AND S. LIN [1970], An efficient heuristic procedure for partitioning graphs, *The Bell System Technical Journal* 49, 291–307.

KHANNA, S., R. MOTWANI, M. SUDAN, AND U. VAZIRANI [1998], On syntactic versus computational views of approximability, *SIAM Journal on Computing* 28, 164–191.

KIRKPATRICK, S., C.D. GELATT, JR., AND M.P. VECCHI [1983], Optimization by simulated annealing, *Science* 220, 671–680.

KLAUCK, H. [1996], On the hardness of global and local approximation, *Proceedings of the 5th Scandinavian Workshop on Algorithm Theory*, Reykjavík, Iceland, 88–99.

KNUTH, D.E. [1998], *The Art of Computer Programming; Volume 3, Sorting and Searching*, Addison-Wesley.

KOLONKO, M. [1999], Some new results on simulated annealing applied to the job shop scheduling problem, *European Journal of Operational Research* 113, 123–136.

KORTE, B., AND J. VYGEN [2002], *Combinatorial Optimization: Theory and Algorithms*, Springer.

KORUPOLU, M., C.G. PLAXTON, AND R. RAJARAMAN [2000], Analysis of a local search heuristic for facility location problems, *Journal of Algorithms* 37, 146–188.

KRENTEL, M.W. [1989], Structure in locally optimal solutions, *Proceedings of the 30th Annual Symposium on Foundations of Computer Science*, Los Alamitos, CA, 216–221.

LAARHOVEN, P.J.M. VAN [1988], *Theoretical and Computational Aspects of Simulated Annealing*, Ph.D. thesis, Eindhoven University of Technology, The Netherlands.

LAARHOVEN, P.J.M. VAN, E.H.L. AARTS, AND J.K. LENSTRA [1992], Job shop scheduling by simulated annealing, *Operations Research* 40, 113–125.

LANGSTON, M.A. [1982], Improved 0/1-interchange scheduling, *BIT* 22, 282–290.

LAWLER, E.L. [1976], *Combinatorial Optimization: Networks and Matroids*, Holt, Rinehart, and Winston.

LAWLER, E.L., J.K. LENSTRA, A.H.G. RINNOOY KAN, AND D.B. SHMOYS (eds.) [1985], *The Traveling Salesman Problem*, Wiley.

LEEUWEN, J. VAN, AND A.A. SCHOONE [1981], Untangling a traveling salesman tour in the plan, *Proceedings of the 7th Conference on Graph Theoretic Concepts in Computer Science*, München, Germany, 87–98.

LENSTRA, J.K., A.H.G. RINNOOY KAN, AND P. BRUCKER [1977], Complexity of machine scheduling problems, *Annals of Discrete Mathematics* 1, 343–362.

LEWIS, J.M., AND M. YANNAKAKIS [1980], The node-deletion problem for hereditary properties is NP-complete, *Journal of Computer and System Sciences* 20, 219–230.

LIN, S. [1965], Computer solutions of the traveling salesman problem, *Bell System Technical Journal* 44, 2245–2269.

LIN, S., AND B.W. KERNIGHAN [1973], An efficient heuristic algorithm for the traveling-salesman problem, *Operations Research* 21, 498–516.

LORENCO, H.R., O. MARTIN, AND T. STÜTZLE [2003], Iterated local search, in: F. Glover and G. Kochenberger (eds.), *Handbook of Metaheuristics*, Kluwer Academic Publishers, 321–353.

LU, H., AND R. RAVI [1992], The power of local optimization: approximation algorithms for maximum leaf spanning tree, *Proceedings of the 30th Annual Allerton Conference on Communication, Control, and Computing*, Urbana, IL, 533–542.

LUECKER, G. [1976], unpublished manuscript, Department of Computer Science, Princeton University, Princeton, NJ.

LUNDY, M., AND A. MEES [1986], Convergence of an annealing algorithm, *Mathematical Programming* **34**, 111–124.

MATSUO, H., C.J. SUH, AND R.S. SULLIVAN [1988], A controlled search simulated annealing method for the general job shop scheduling problem, Working paper 03-04-88, Graduate School of Business, University of Texas, Austin, TX.

METROPOLIS, M., A. ROSENBLUTH, M. ROSENBLUTH, A. TELLER, AND E. TELLER [1953], Equation of state calculations by fast computing machines, *Journal of Chemical Physics* **21**, 1087–1092.

MICHALEWICZ, Z. [1992], *Genetic Algorithms + Data Structures = Evolution Programs*, Springer.

MICHALEWICZ, Z., AND D.B. FOGEL [2000], *How to Solve It: Modern Heuristics*, Springer.

MITRA, D., F. ROMEO, AND A.L. SANGIOVANNI-VINCENTELLI [1986], Convergence and finite-time behavior of simulated annealing, *Advances in Applied Probability* **18**, 747–771.

MOTWANI, R., AND P. RAGHAVAN [1995], *Randomized Algorithms*, Cambridge University Press.

MÜHLENBEIN, H., M. GORGES-SCHLEUTER, AND O. KRÄMER [1988], Evolution algorithms in combinatorial optimization, *Parallel Coputing* **7**, 65–85.

NEMHAUSER, G.L., AND L.A. WOLSEY [1988], *Integer and Combinatorial Optimization*, Wiley.

NICHOLSON, T.A.J. [1971], A method for optimizing permutation problems and its industrial appications, in: D.J.A. Welsh (ed.), *Combinatorial Mathematics and Its Applications*, Academic Press, 201–217.

NOLTE, A., AND R. SCHRADER [1996], Simulated annealing and its problems to color graphs, *Proceedings of the 4th Annual European Symposium on Algorithms*, Barcelona, Spain, 138–151.

NOLTE, A., AND R. SCHRADER [1997], Coloring in sublinear time, *Proceedings of the 5th Annual European Symposium on Algorithms*, Graz, Austria, 388–401.

NOLTE, A., AND R. SCHRADER [2000], A note on the finite time behavior of simulated annealing, *Mathematics of Operations Research* **25**, 476–484.

O'HEIGEARTAIGH, M., J.K. LENSTRA, AND A.H.G. RINNOOY KAN (eds.) [1985], *Combinatorial Optimization: Annotated Bibliographies*, Wiley.

OR, I. [1976], *Traveling salesman-type combinatorial problems and their relation to the logistics of regional blood banking*, Ph.D. thesis, Northwestern University, Evanston, IL.

ORLIN, J.B., A.P. PUNNEN, AND A.S. SCHULZ [2004], Approximate local search in combinatorial optimization, *SIAM Journal on Computing* **33**, 1201–1214.

OROSZ, J.E., AND S.H. JACOBSON [2002], Finite-time performance analysis of static simulated annealing algorithms, *Computational Optimization and Applications* **21**, 21–53.

OSBORNE, L.J., AND B.E. GILLETT [1991], A comparison of two simulated annealing algorithms applied to the directed Steiner problem in networks, *ORSA Journal on Computing* **3**, 213–225.

OSMAN, I.H., AND G. LAPORTE [1996], Metaheuristics: A bibliography, *Annals of Operations Research* **63**, 513–623.

PAPADIMITRIOU, C.H. [1977], The Euclidean traveling salesman problem is NP-complete, *Theoretical Computer Science* **4**, 237–244.

PAPADIMITRIOU, C.H. [1992], The complexity of the Lin-Kernighan heuristic for the traveling salesman problem, *SIAM Journal on Computing* **21**, 450–465.

PAPADIMITRIOU, C.H. [1994], *Computational Complexity*, Addison-Wesley.

PAPADIMITRIOU, C.H., A.A. SCHÄFFER, AND M. YANNAKAKIS [1990], On the complexity of local search, *Proceedings of the 22^{nd} ACM Symposium on Theory of Computing*, Baltimore, MD, 438–445.

PAPADIMITRIOU, C.H., AND K. STEIGLITZ [1977], On the complexity of local search for the traveling salesman problem, *SIAM Journal on Computing* **6**, 76–83.

PAPADIMITRIOU, C.H., AND K. STEIGLITZ [1978], Some examples of difficult traveling salesman problems, *Operations Research* **26**, 434–443.

PAPADIMITRIOU, C.H., AND K. STEIGLITZ [1982], *Combinatorial Optimization*, Prentice-Hall.

PAPADIMITRIOU, C.H., AND S. VEMPALA [2000], On the approximability of the traveling salesman problem, *Proceedings of the 32^{nd} ACM-SIAM Symposium on Theory of Computing*, Portland, OR, 126–133, The paper contains errors that affected the results. A corrected version of the paper can be found on the authors' homepages.

PINEDO, M.L. [1995], *Scheduling: Theory, Algorithms, and Systems*, Prentice Hall.

PITSOULIS, L.S., AND M.G.C. RESENDE [2002], Greedy randomized adaptive search procedures, in: P.M. Pardalos and M.G.C. Resende (eds.), *Handbook of Applied Optimization*, Oxford University Press, 168–183.

POLJAK, S. [1995], Integer linear programs and local search for max-cut, *SIAM Journal on Computing* **24**, 822–839.

PRIM, R.C. [1957], Shortest connection networks and some generalizations, *Bell System Technical Journal* **36**, 1389–1401.

PUNNEN, A.P., M. MARGOT, AND S.N. KABADI [2003], TSP heuristics: Domination analysis and complexity, *Algorithmica* **35**, 111–127.

RAGHAVACHARI, B. [1997], Algorithms for finding low degree structures, in: D.S. Hochbaum (ed.), *Approximation Algorithms for NP-hard Problems*, PWS Publishing Company, 266–295.

REEVES, C.R. (ed.) [1993], *Modern Heuristic Techniques for Combinatorial Problems*, Wiley.

REINELT, G. (ed.) [1994], *The Traveling Salesman: Computational Solutions for TSP Applications*, Springer.

RIBEIRO, C.C., AND P. HANSEN (eds.) [2002], *Essays and Surveys on Metaheuristics*, Kluwer Academic Publishers.

ROBERTSON, N., D. SANDERS, P. SEYMOUR, AND R. THOMAS [1997], The four-colour theorem, *Journal of Combinatorial Theory, Series B* **70**, 2–44.

ROBINS, G., AND A. ZELIKOVSKY [2000], Improved Steiner tree approximation in graphs, *Proceedings of the 10^{th} ACM-SIAM Symposium on Discrete Algorithms*, San Francisco, CA, 770–779.

ROSENTHAL, R.W. [1973], A class of games possessing pure-strategy Nash equilibria, *International Journal of Game Theory* **2**, 65–67.

ROY, B., AND B. SUSSMANN [1964], *Les Problèmes d'ordonnancement avec contraintes disjonctives*, Technical Report Note DS No. 9 bis, SEMA, Montrouge, France.

SAATY, T.L., AND P.C. KAINEN [1977], *The Four-Color Problem: Assaults and Conquest*, McGraw-Hill.

SAHNI, S., AND T. GONZALEZ [1976], P-complete approximation problems, *Journal of the Association for Computing Machinery* **23**, 555–565.

SALAMON, P., P. SIBANI, AND R. FROST [2002], *Facts, Conjectures, and Improvements for Simulated Annealing*, SIAM Monographs on Mathematical Modeling and Computation.

SASAKI, G., AND B. HAJEK [1988], The time complexity of maximum matching by simulated annealing, *Journal of the ACM* **35**, 387–403.

SCHÄFFER, A.A., AND M. YANNAKAKIS [1991], Simple local search problems that are hard to solve, *SIAM Journal on Computing* **20**, 56–87.

SCHILHAM, R.M.F. [2001], *Commonalities in Local Search*, Ph.D. thesis, Eindhoven University of Technology, The Netherlands.

SCHRIJVER, A. [2003], *Combinatorial Optimization*, Springer.

SCHULZ, A.S., R. WEISMANTEL, AND G.M. ZIEGLER [1995], 0/1-integer programming: Optimization and augmentation are equivalent, *Proceedings of the 3rd Annual European Symposium on Algorithms*, Corfu, Greece, 473–483.

SCHUURMAN, P., AND T. VREDEVELD [2001], Performance guarantees of local search for multiprocessor scheduling, *Proceedings of the 8th Conference on Integer Programming and Combinatorial Optimization*, Utrecht, The Netherlands, 370–382.

SENETA, E. [1981], *Non-negative Matrices and Markov Chains*, Springer.

SHIMOZONO, S. [1997], Finding optimal subgraphs by local search, *Theoretical Computer Science* **172**, 265–271.

SIPSER, M. [1997], *Introduction to the Theory of Computation*, PWS Publishing Company.

SMITH, W.E. [1956], Various optimizers for single-stage production, *Naval Research Logistics Quarterly* **3**, 59–66.

SORKIN, G.B. [1991], Efficient simulated annealing on fractal energy landscapes, *Algorithmica* **6**, 367–418.

STADLER, P.F., AND W. SCHNABL [1992], The landscape of the traveling salesman problem, *Physics Letters A* **161**, 337–344.

STEINHÖFEL, K., A.A. ALBRECHT, AND C.K. WONG [1998], On various cooling schedules for simulated annealing applied to the job shop problem, *Proceedings of the 2nd International Workshop on Randomized and Approximation Techniques in Computer Science*, Barcelona, Spain, 260–279.

TOVEY, C. [1997], Local improvement on discrete structures, in: E.H.L. Aarts and J.K. Lenstra (eds.), *Local Search in Combinatorial Optimization*, Wiley, 57–89.

ULDER, N.L.J., E.H.L. AARTS, H.J. BANDELT, P.J.M. VAN LAARHOVEN, AND E. PESCH [1991], Genetic local search algorithms for the traveling salesman problem, *Proceedings of the 1st Workshop on Parallel Problem Solving from Nature*, Dortmund, Germany, 109–116.

VREDEVELD, T. [2002], *Combinatorial Approximation Algorithms: Guaranteed Versus Experimental Performance*, Ph.D. thesis, Eindhoven University of Technology, The Netherlands.

WASSERMAN, P.D. [1989], *Neural Computing: Theory and Practice*, Van Nostrand Reinhold.

WEGENER, I. [2005], Simulated annealing beats Metropolis in combinatorial optimization, *Proceedings of the 32nd International Colloquium on Automata, Languages and Programming*, Lisbon, Portugal, 589–601.

WILSON, R.J., AND J.J. WATKINS [1990], *Graphs: An Introductory Approach*, Wiley.

WINKER, P. [2001], *Optimization Heuristics in Econometrics: Applications of Threshold Accepting*, Wiley.

YANNAKAKIS, M. [1997], Computational complexity, in: E.H.L. Aarts and J.K. Lenstra (eds.), *Local Search in Combinatorial Optimization*, Wiley, 19–55.

Author Index

Subject Index

Monographs in Theoretical Computer Science · An EATCS Series

Texts in Theoretical Computer Science · An EATCS Series